A Fuller Explanation

The Synergetic Geometry of R. Buckminster Fuller

Design Science Collection

Series Editor
Arthur L. Loeb
Department of Visual and Environmental Studies
Harvard University

Amy C. Edmondson   *A Fuller Explanation: The Synergetic*
                    Geometry of R. Buckminster Fuller, 1987

In Preparation

Marjorie Senechal   *Shaping Space: A Polyhedral Approach*
  and George Fleck
  (Eds.)
Arthur L. Loeb   *Concepts and Images*

Amy C. Edmondson

A Fuller Explanation
The Synergetic Geometry of
R. Buckminster Fuller

A *Pro Scientia Viva* Title

Birkhäuser
Boston · Basel · Stuttgart

Amy C. Edmondson
A Fuller Explanation
The Synergetic Geometry of R. Buckminster Fuller

Coden: DSCOED

First Printing, 1987

Library of Congress Cataloging in Publication Data
Edmondson, Amy C.
 A Fuller explanation. The synergetic geometry of R. Buckminster Fuller.
 (Design science collection)
 "A Pro scientia viva title."
 Bibliography: p.
 Includes index.
 1. System theory. 2. Thought and thinking.
3. Mathematics—Philosophy. 4. Geometry—Philosophy.
5. Fuller, R. Buckminster (Richard Buckminster).
1895– . I. Title. II. Series.
Q295.E33 1986   003   86-14791

CIP-Kurztitelaufnahme der Deutschen Bibliothek
Edmondson, Amy C.:
A Fuller explanation : the synerget. geometry of
R. Buckminster Fuller / Amy C. Edmondson—1.
print. —Boston ; Basel ; Stuttgart : Birkhäuser,
1986.
 (Design science collection) (A pro scientia
 viva title)
 ISBN 3-7643-3338-3 (Basel. . .)
 ISBN 0-8176-3338-3 (Boston)

Frontispiece photograph courtesy of Phil Haggerty.

ISBN 0-8176-3338-3
    3-7643-3338-3

Typeset by Science Typographers, Inc., Medford, New York.
Printed and bound by R. R. Donnelley & Sons, Harrisonburg, Virginia.
Manufactured in the United States of America.

To my parents
Mary Dillon Edmondson and *Robert Joseph Edmondson*

Contents

Series Editor's Foreword

In a broad sense Design Science is the grammar of a language of images rather than of words. Modern communication techniques enable us to transmit and reconstitute images without the need of knowing a specific verbal sequential language such as the Morse code or Hungarian. International traffic signs use international image symbols which are not specific to any particular verbal language. An image language differs from a verbal one in that the latter uses a linear string of symbols, whereas the former is multidimensional.

Architectural renderings commonly show projections onto three mutually perpendicular planes, or consist of cross sections at different altitudes representing a stack of floor plans. Such renderings make it difficult to imagine buildings containing ramps and other features which disguise the separation between floors; consequently, they limit the creativity of the architect. Analogously, we tend to analyze natural structures as if nature had used similar stacked renderings, rather than, for instance, a system of packed spheres, with the result that we fail to perceive the system of organization determining the form of such structures.

Perception is a complex process. Our senses record; they are analogous to audio or video devices. We cannot claim, however, that such devices perceive. Perception involves more than meets the eye: it involves processing and organization of recorded data. When we classify an object, we actually name an abstract concept: such words as *octahedron, collage, tessellation, dome*; each designates a wide variety of objects sharing certain characteristics. When we devise ways of transforming an octahedron, or determine whether a given shape will tessellate the plane, we make use of these characteristics, which constitute the grammar of structure.

The Design Science Collection concerns itself with various aspects of this grammar. The basic parameters of structure, such as symmetry, connectivity, stability, shape, color, size, recur throughout these volumes. Their interactions are complex; together they generate such concepts as Fuller's and Snelson's tensegrity, Lois Swirnoff's modulation of surface through color, self-reference in the work of M. C. Escher, or the synergetic stability of ganged unstable polyhedra. All of these occupy some of the professionals concerned with the complexity of the space in which we live, and which we shape. The Design Science Collection is intended to inform a reasonably well-educated but not highly specialized audience of these professional activities, and particularly to illustrate and to stimulate the interaction between the various disciplines involved in the exploration of our own three-dimensional, and in some instances more-dimensional, spaces.

When R. Buckminster Fuller recalled his days as a schoolboy in Milton, Massachusetts, he related how his mathematics teacher would introduce two-dimensional surfaces by placing lines of zero thickness side by side; young Buckminster used to wonder how one could create a finite surface out of nothing. Similarly, he could not accept the stacking of planes of zero thickness to create volumes. Intuitively, he sensed that areas and volumes are as different from each other as are forces and velocities: one cannot mix quantities of different dimensionality. Accordingly, Fuller learned to compare three-dimensional objects with each other, and hence to add, subtract, and transform them from and into each other rather than creating them out of objects of lower dimensionality. In doing so he came to discard the conventional orthogonal system which has blinded architects as well as solid-state scientists, and followed natural structure in designing his stable light-weight structures.

Two days before Harvard Commencement in 1983, Amy Edmondson called me from Buckminster Fuller's office in Philadelphia, saying that Fuller had decided at the last moment to attend the Commencement exercises, and wondered whether I might still be free to have dinner with them the following evening. Amy had graduated from Harvard with combined honors in Applied Science and in Visual and Environmental Studies, and had been working for Fuller since then. At dinner we planned a working session in August at Fuller's island off the Maine coast. Unfortunately that Commencement turned out to be Buckminster's last, and when I saw Amy again it was at the combined service in memory of Buckminster and Anne Fuller. We decided right then and there that the best tribute would be a volume aiming at translating Buckminster Fuller's ideas and idiom into a language more accessible to the lay audience and more acceptable to the scientist.

Amy Edmondson has succeeded admirably in conveying to us not just the idiom but also the atmosphere of Fuller's "office." There were no professional draftsmen, for the staff was minimal. We believe that the sense of a direct link to the Fuller office would be enhanced by reproducing Edmondson's own illustrations directly, just as she would have produced them there.

With *A Fuller Explanation* we initiate the Design Science Collection, an exploration of three-dimensional space from the varied perspectives of the designer, artist, and scientist. Through this series we hope to extend the repertoire of the former to professions by using natural structure as an example, and to demonstrate the role of esthetic sensibility and an intuitive approach in the solution of scientific problems.

ARTHUR L. LOEB

Cambridge, Massachusetts

Preface

Buckminster Fuller has been alternately hailed as the most innovative thinker of our time and dismissed as an incomprehensible maverick, but there is a consistent thread running through all the wildly disparate reactions. One point about which there is little disagreement is the difficulty of understanding Bucky. "It was great! What did he say?" is the oft-repeated joke, describing the reaction of a typical enraptured listener after one of Fuller's lectures.

Not surprisingly then, Fuller's mathematical writing has not attracted a mass audience. Rather, synergetics has become a sort of Fuller proving ground, into which only a few scientific-minded types dare to venture. "Oh, I'll never be able to understand it then" has been the response of countless people upon learning that the subject of my book is synergetics. This reaction would have saddened Bucky immensely: he was so sure his geometry was appropriate for five-year-olds! However, such shyness is understandable; deciphering Fuller's two volumes, *Synergetics* and *Synergetics 2*, requires a sizable commitment of time and patience from even the most dedicated reader. Study groups have gone a long way toward helping individuals unravel the idiosyncratic, hyphenated prose of these two works, but the task, still arduous, is not for everyone. However, as those who dared it will have discovered, the major concepts presented in Fuller's intimidating books are not inherently difficult, and much of synergetics can be explained in simple, familiar terms. That is the purpose of *A Fuller Explanation*.

Synergetics, in the broadest terms, is the study of spatial complexity, and as such is an inherently comprehensive discipline. Designers, architects, and scientists can easily find applications of this study in their work; however, the larger significance of Fuller's geometry may be less visible. Experience with synergetics encourages a new way of approaching and solving problems. Its emphasis on visual and spatial phenomena combined with Fuller's holistic approach fosters the kind of lateral thinking which so often leads to creative breakthroughs.

A Fuller Explanation is geared to readers with no mathematical background, but of course it can be read at many levels. Even if one is familiar with some of the concepts, Fuller's unique interpretation and development of them will be enlightening. This book should appeal to anyone interested in patterns and design and how things work.

Synergetics is also fascinating as a reflection of Fuller himself; his wide-eyed appreciation of nature and human invention alike exudes from his expression of these geometric concepts. The primary purpose of this volume is thus to present the nuts and bolts of synergetics, the tools with which to continue exploring this discipline; but almost as importantly, I hope to convey the spirit of Fuller's inquiry into the organizing principles of nature.

When Bucky Fuller looked around, he saw, not trees and roads and butterflies, but a miraculous web of interacting patterns. As he describes these patterns, using his peculiar blend of antiquated phrases and electronic-age jargon, one cannot help concluding that no child was ever as startled as the young Bucky to discover that the world is not what it appears, that the apparently solid and lifeless rock is a bundle of energetic atomic activity. He never lost that awe. Synergetics is his attempt to give some of it away.

I have included (along with explanations and definitions) many of Fuller's own descriptions and invented terms. Although I set out to interpret and explain Bucky in ordinary language—a task I found increasingly tricky as I became more deeply involved in the project—I began to sense a deep appropriateness to Bucky's peculiar phraseology. Many passages that seem convoluted at first reading later seem to express his meaning more precisely than could any substitutes. As I quote Fuller often in this book, readers will be able to judge for themselves.

Acknowledgments

Arthur Loeb first introduced me to the intricate structure of ordered space through his wonderful course at Harvard University. As editor of this book, he generously resumes his instructive role—seven years later, long after his responsibility to a former student has expired. For this I am deeply grateful. Dr. Loeb has perfected the art of gentle criticism, and it was an extraordinary privilege to have had this exposure to his kindness and wisdom.

I am greatly indebted to Bucky's family, especially Allegra Fuller Snyder and Jaime Snyder, for their enthusiastic endorsement of my decision to write this book; their commitment to making available information about Bucky's work and life has enabled *A Fuller Explanation* to exist. I also want to thank John Ferry at the Buckminster Fuller Institute for his assistance in locating photographs.

I am grateful to many people for their encouragement and inspiration, but above all to my parents, Bob and Mary Edmondson, whose support of this project was indispensable.

I want to thank Martha Lerski for her eager and skillful editing; her help was invaluable. Also, I profited enormously from the insights of many who have attended synergetics workshops, but to none am I more indebted than to Darrell Mullis, Tony Perry, Martha Stires Perry, and Carrie Fisher.

My husband, Mark Carpenter, provided the constant sustenance that nurtured both this book and its author throughout the writing process. He contributed many of the drawings, hours of editing, and most importantly his inexhaustible intellectual curiosity.

Finally, Bucky was and remains an endless source of inspiration. He was the best teacher and friend any 21-year-old just entering the "real world" could have had. It would be redundant to try to express my gratitude for this experience here, for that is what I hope this book will do.

AMY C. EDMONDSON

New York, New York

Note to Readers

Quotations from Fuller's *Synergetics: Explorations in the Geometry of Thinking*, (which is organized into numbered sections and subsections in such a way that each paragraph is given its own reference number) are followed by the numerical section reference in parentheses.

Passages from *Synergetics 2: Further Explorations in the Geometry of Thinking* will be referenced likewise, with the addition of the letter "b" after the section number.

Photograph courtesy of ARGO Construction, Inc., Montreal, Quebec, Canada.

Introduction

The scene is Montreal, 1967: travelers from around the world emerge from a subway station at the Expo site, and catch their first glimpse of an enormous transparent bubble. Looking and exclaiming, they gravitate toward this strange monument, which is the United States Pavilion, and few notice the stocky white-haired old man, straining his slightly deaf ears to glean their reactions. Buckminster Fuller, playing the disinterested bystander, along with Anne Hewlett Fuller, his wife of exactly fifty years, is a triumphant eavesdropper; the candid observers have enthusiastically approved his design. Fifteen years later, he recalls that summer morning with a playful grin, clearly enjoying the memory of his short stint as detective, and I can almost see him there, standing next to Anne, silently sharing the knowledge that the years of perserverence—ignoring skepticism and often decidedly harsh disapproval of his mathematical work—were vindicated.

Bucky is such a gifted story teller that I also imagine I can see the huge geodesic sphere reflecting the intense summer sun, and it looks more like one of nature's creations than architecture. But it is steel, Plexiglas and human ingenuity that have created this glittering membrane, which was, in 1967, the world's largest dome, spanning 250 feet and reaching an altitude equal to that of a twenty-story building, without any interior support.

More than just the millions who visited Expo '67 have admired this architectural feat, and humanity has found countless other uses for the geodesic dome, as evidenced by the 100,000 such structures of various materials and sizes that are sprinkled around the globe. However, the "synergetic geometry," which lies behind Fuller's remarkable design, has remained almost completely obscure.

The goal of this book is to catalyze a process which I hope will continue and expand on its own: to rescue Fuller's fascinating material from its unfortunate obscurity.

A. C. E.

Experience has shown repeatedly that a mathematical theory with a rich internal structure generally turns out to have significant implications for the understanding of the real world, often in ways no one could have envisioned before the theory was developed.

William P. Thurston and Jeffrey R. Weeks
The Mathematics of Three-Dimensional Manifolds
Scientific American, July 1984.

1

Return to Modelability[1]

Synergetic geometry is the product of a mind as comfortable with mathematical precision as with the intuitive leaps associated with visual and spatial conceptualizing. Buckminster Fuller was guided predominantly by intuition throughout his 87 years; nonetheless, he was entirely at ease with the painstaking exactitude of numerical calculation—such as that required in the development of the geodesic dome in the early 1940s. Years before the pocket calculator, he produced volumes of intricate trigonometric solutions, manipulating eight-digit numbers with the patience and precision of a monk. However, the peculiar language of Fuller's mathematical writings quickly betrays the intuitive influence and all but conceals that of the hard-nosed engineer. Buckminster Fuller was both the pragmatic Yankee mechanic and the enigmatic mystic, and synergetics is the product of that combination.

Above all, he was driven by curiosity—and found nature a far more compelling teacher than the textbooks in his Milton, Massachusetts, schoolhouse. Frustrated by the apparent lack of a connection between conventional mathematics and reality, young "Bucky" Fuller adopted his own approach. The resulting self-directed exploration into pattern and structure became the most powerful influence in his remarkable career as inventor, architect, engineer, and philosopher, and produced a geometrical system that provides useful background for problem-solving of any kind.

Synergetics is the discipline hiding behind Fuller's fantastic visions of a sustainable future. These reliable patterns were the source of his unshakable confidence in his design-science philosophy, which—in short—upholds that innovative application of the principles governing nature's behavior can insure ample life support for all humanity. While many people around the world have been exposed to Fuller's ideas and inventions, few have understood or even been aware of the

[1]*See* Notes, pp. 289ff.

mathematical principles underlying the elegant efficiency of structures such as the Octet Truss and geodesic dome. Happily, these principles are easily accessible once you get into the spirit of Fuller's approach: synergetics is a "hands-on" branch of mathematics.

However, listening to one of Fuller's all-encompassing lectures, you might wonder when the "hands-on" part begins. Tangibility is not a prominent feature in his spell-binding discourse, the subject of which is no less than "humans in universe." He challenges, in the course of a few hours, age-old assumptions about our lives and institutions, asking us to reconsider the most commonplace aspects of experience. Some of his observations are stated so simply, you may find yourself wondering, "Why haven't I thought about that before"? For example:

> How many of you have said to your children, "darling, look at the beautiful sun going down"? [A show of many hands.] "Well, we've known for five-hundred years that the sun *isn't* going down, and yet we consider it practical to keep on lying to our children!

Or:

> When I was born in 1895, reality was everything you could see, smell, touch and hear. The world was thought to be absolutely self-evident. When I was three years old, the electron was discovered. That was the first invisible. It didn't get in any of the newspapers; (nobody thought *that* would be important!) Today 99.99% of everything that affects our lives cannot be detected by the human senses. We live in a world of invisibles.

And later, he takes his keys out of his pocket and carelessly tosses them in the air; gravity takes care of the landing.

> Nature doesn't have to have department meetings to decide what to do with those keys, or how to grow a turnip. She knows *just* what to do. It must be that nature has only one department, one coordinating system.[2]

These simple truths each relate to different aspects of synergetic geometry. But for all his lighter anecdotes, Fuller's underlying message could not have been more serious:

> The fact that 99 percent of humanity does not understand nature is the prime reason for humanity's failure to exercise its option to attain universally sustainable physical success on this planet. The prime barrier to humanity's discovery and comprehension of nature is the obscurity of the mathematical language of science. Fortunately, however, nature is *not* using the strictly imaginary, awkward, and unrealistic coordinate system adopted by and taught by present-day academic science. (000.125b)

Nature is instead using the principles embodied in synergetics, which thus provides the way to eradicate this lethal ignorance. Claiming to have discovered no less than the mathematical system that describes the coordination of physical and metaphysical phenomena alike—that is, of both energy and thought—Fuller was urgent in his insistence that we study these principles:

I am confident that humanity's survival depends on all of our willingness to comprehend feelingly the way nature works.[2]

From Geometry to Geodesics: A Personal Perspective

What college student (or human being for that matter) would not be overjoyed to receive an invitation to work on "ever-more relevant affairs" from the person she most admires? That is precisely what I found in my mailbox in 1980 when Buckminster Fuller actually answered my letter, the timid plea of an undergraduate: "What can people *do* toward furthering your vision of making this planet work for everyone? And where can I *apply* the experience of having studied synergetics"? I was later to understand that Fuller's responding to an undergraduate's letter was not a miracle but instead revealed his profound trust in the integrity and capability of human beings—and especially of youth. His action was typical of his life and work, which relied heavily on intuition, with a powerful faith in the willingness of others to apply their minds as diligently and joyfully as he applied his. We have to take Fuller at his word when he claims to be not a genius but an "average healthy human being" who exercised his option to think. He embraced that potential in all of us.

I was introduced to the intricate discipline of geometry in a Harvard course "Synergetics: the structure of ordered space" taught by the editor of this series, Arthur L. Loeb and I had been fascinated by this material for a couple of years. Reconciled to its obscurity, I was enchanted by the perfection and complexity of this body of geometric knowledge, which was all but completely hidden from popular awareness. In those days Loeb's course was similarly hidden, a bizarre option within the two-inch-thick course catalog, taught in a sequestered attic in Sever Hall where one would never wander accidentally. My peers had no doubt that there was a reason for that. In fact, my academic pursuits were perceived by most as an irreverent cross between kindergarten games and mathematical torture. My roommates, forever tripping over cardboard tetrahedra and unsuccessful tensegrity wheelbarrows while gingerly avoiding small deposits of Elmer's glue, were tolerantly confused. I can't blame my classmates for their bemused head-shaking; I had trouble taking my*self* seriously. Mathematical elegance aside, I felt deep down that I had chosen an unbelievably fascinating road to nowhere, a choice that would no doubt ultimately bar me from all chance of meaningful participation in human affairs. But still I was trapped—like an addict immune from better judgment—in my polyhedral playpen.

Then one February evening, I heard Fuller speak at the Massachusetts Institute of Technology. It's easy to understand how pivotal the experience could be: in love with geometry but distressed by its nonapplicability, I heard Bucky Fuller that night spin out—in an omnidirectional web of ideas, predictions, and obscure but brilliantly juxtaposed facts—an unfamiliar version of world history in which synergetic geometry (and other aspects of comprehensive thinking) somehow played a crucial role in rescuing humanity from its current crisis of squandering vast resources in an unwinnable arms race. We are suddenly at a turning point in history at which it is possible to provide adequate life support for everyone, declared Bucky. Malthus is obsolete. (He didn't know about alloys.[3]) There is no such thing as a straight line, the sun does not go down, and it is time we updated our language.

A funnier, more serious, more mesmerizing discourse I have never heard. I walked—no, skipped—back home down Massachusetts Avenue: not that I could have told you exactly how it worked, this planetary success, but I was sold. My geometry had relevance! I worked harder than ever.

I still can't explain exactly how synergetics is going to turn the world around, but I have found at least that I can explain synergetics. My hope is that if enough other people become aware of these principles, the missing pieces will ultimately come together. So far, this has been a valid working hypothesis. In giving lectures and workshops to clarify Fuller's material, I have met people who found significant applications in their own work. Synergetics has provided both useful models to elucidate scientific phenomena and methods of solving structural problems. Examples of both aspects will be cited throughout the book; see especially the end of Chapter 15, "Case In Point: Donald Ingber" which can easily be read independently of the rest of the book.

After Fuller's lecture, the next step was clear: read *Synergetics*. If I expected easy answers, I was in for a surprise. Fuller's ambiguous writing called for considerable interpretation. With the patient guidance of Arthur Loeb, I struggled through Fuller's massive text and learned that truth was far more elusive than he had made it sound that night at MIT. But the geometry was no less seductive, and ultimately I decided to risk a thirteen-cent stamp.

And then his unexpected letter arrived—in response to my earnest but decidedly indirect questions. Even a photocopied list of organizations would have been received with joy. How it was that my letter filtered through the procedural maze that lay between Buckminster Fuller and the formidable stack of mail that was opened and sorted by various trusted assistants every day, I'll never know.

The signature was real:

Dear Amy Edmondson:
...I would like to take advantage of your offer to come and work with me. ...I am busier and busier with ever more relevant affairs.

Warmly, Faithfully,
Buckminster Fuller

Ever more relevant affairs! A college student's dream and—she is convinced after three years of first-hand acquaintance—an accurate description of Fuller's experience. Even with a healthy dose of skepticism about some aspects of his philosophy, one could not help being stunned by his tireless enthusiasm for work. At four times my age, he was awake and working before I arrived and long after I had crept home to bed exhausted. The secret of this energy was his conviction that humanity had a viable option of designing an unprecedentedly successful environment aboard "Spaceship Earth"[4], and that his work just might play an important part in that. I found that the more deeply involved in the actual work I became (in my case, calculations and drawings for Fuller's engineering projects, including the progressive refinement of geodesic designs), the more impressed I was by the scope of his vision. Gradually, as is generally the case, naïve youthful worship disappeared. But as is *not* generally the case, its place was taken by a new deep respect for the mind of my friend, Bucky. (It would be unimaginable for anyone who ever spent more than an hour with Bucky to call him anything else, with the sole exception of his friend and colleague, Arthur Loeb, who calls him "Buckminster.")

Bucky was extraordinarily generous with his time—perhaps due to an uncontrollable urge to teach—and treated every listener as an intellectual equal. This might be called a brilliant teaching strategy, except that it was utterly spontaneous. One of the most important lessons of my three-year experience was the difference between Bucky on the other side of his desk—spontaneously lapsing into simple clear explanations as a result of the catalyst of a pair of expectant human eyes which would cloud into a worried frown when lost—and Buckminster Fuller's dense polysyllabic prose in the 800 pages of *Synergetics*. I became accustomed to translating the Fullerese into lay English for various befuddled readers who went so far as to call the office for help.

This is not a book about those three years; it is about synergetic geometry. Here I have only tried to give you some of the background that has led me to attempt to explain what is in many ways unexplainable, for no one can speak for Bucky Fuller but himself. The goal of this volume is to help readers get through the barriers imposed by Fuller's idiosyncratic use of language, and to introduce the major concepts of synergetics in an accessible format. The next steps are up to all of us.

Operational Mathematics

We can imagine the young Bucky, an enthusiastic misfit sensing that he is alone in his skepticism about the fundamental premises of geometry. ("Does *no* one see what I see? Does no one else sense the terrible problems that lie ahead if we follow these absurd premises to their logical ends?") While his grade-school classmates were apparently content to go right along with the teacher's strange games without complaint, Bucky was astonished by the implausible new concepts.

Bucky would tell us that he tried, constantly, to accept the rules—be a good student, make his family proud, submit to and even excel at the illogical activities—but somehow his efforts at model behavior were always thwarted. He just couldn't help pointing out that the teacher's "straight line" was not at all straight, but rather slightly curved and definitely fragmented. Perplexed by her lack of accuracy, Bucky saw a trail of powdery chalk dust left on the blackboard, a trace of the motion of her hand, and it seemed quite unlike her words.

Bucky was virtually blind until he got his first eyeglasses at the age of five, so he had truly experienced life without this primary sense. Now he was insatiably curious about the visual patterns around him. An "infinite straight line"? He would turn toward the window, thoughtfully pondering where that "infinite line" stopped. "Out the window and over the hill and on and on it goes"; it didn't seem right somehow. Bucky, childishly earnest even in his eighties, would tell this story, explaining that he didn't mean to be fresh, he just couldn't help wondering if that teacher really knew what she was talking about.

Much later, fascinated by Eddington's[5] definition of science as the systematic attempt to set in order the facts of experience, Fuller had a plan. It must be possible to develop a mathematical system consistent with experience. He concluded that humanity had been on the wrong track all these years.

Experimental Evidence

It seemed to Fuller that mathematicians arbitrarily invent impossible concepts, decide rules for their interaction, and then memorize the whole game. But what did he propose as an alternative?

Starting from scratch. Mathematical principles must be derived from experience. Start with real things, observe, record, and then deduce. Working with demonstrable (as opposed to impossible) concepts, the resulting generalizations would reflect and apply to the world in which we live. It seemed highly likely that such an experimental approach would lead to a comprehensive and entirely rational set of principles that represented actual phenomena. Furthermore, Bucky suspected that such an inventory would relate to metaphysical as well as physical structure.

Fuller decided that to begin this process of rethinking mathematics he had to ask some very basic questions. What does exist? What are the characteristics of existence? He proposed that science's understanding of reality should be incorporated into new models to replace the no longer appropriate cubes and other "solids" that had kept mathematicians deliberately divorced from reality since the days of ancient Greece.

To begin with, there are no "solids"; matter consists exclusively of energy. "Things" are actually *events*—transient arrangements of frenetically vibrating atomic motion. It's almost unthinkable, but perhaps if we get our vocabulary and models to be more consistent with an energy-event reality, humanity can be retuned to become comfortable with the invisible discoveries of science. In short, simplifies Fuller, we have the option to tell children the truth about nature in the first place.

What if we do go along with the rule that mathematics cannot define and depend upon a concept that cannot be demonstrated; where does that leave us? Fuller saw inconsistencies even in the notion of an "imaginary straight line," for imagination relies on experience ("image-ination" he would say) to construct its images. Therefore an "operational mathematics" must rely on concepts that correspond to reality.

Bucky pulls us back into the turn-of-the-century schoolhouse of his childhood. The teacher stood at the blackboard, made a little dot, and said, "This is a point; it doesn't exist." ("So she wiped that out.") Then she drew a whole string of them and called it a "line." Having no thickness, it couldn't exist either. Next she made a raft out of these lines and came up with a "plane." "I'm sorry to say it didn't exist either," sighs Bucky. She then stacked them together and got a "cube," and suddenly *that* existed. Telling the story, Bucky scratches his head as if still puzzled seventy years later: "I couldn't believe it; how did she get existence out of nonexistence to the fourth power? So I asked, 'How old is it'? She said, 'don't be naughty.' ... It was an absolute ghost cube."[6]

Instead of a dimensionless "point," Fuller proposes the widely applicable "energy event." Every identifiable experience is an energy event, he summarizes, and many are small enough to be considered "points," such as a small deposit of chalk dust. An aggregate of events too distant to be differentiated from one another can also be treated as a "point." Consider for example a plastic bag of oranges carried by a pedestrian and viewed from the top of the Empire State Building, or a star—consisting of immense numbers of speeding particles—appearing as a tiny dot of negligible size despite having an actual diameter far greater than that of the earth.[7]

The mathematician's "straight line," defined as having length but no width, simply cannot be demonstrated. All physical "lines" upon closer inspection are actually wavelike or fragmented trajectories: even a "line of sight" is a wave phenomenon, insists Fuller; "physics has found no straight lines."[2] But forces exist, and they pull or push in a line, which can be modeled by a vector. We shall explore vectors and their role in synergetics in great detail in later chapters. Finally, the "continuous plane" with no thickness must be replaced by a mesh of energy events interrelated by fine networks of tiny vectors. To Fuller, these adjustments were crucial, for mathematicians' games with continuous planes and sizeless points were ultimately irrelevant diversions; however enticing the intellectual pursuit might be, they do not lead to a better understanding of how nature works.

The essential nature of the above revolution is semantic and can easily seem trivial. The difficulty in evaluating the impact of such changes lies in the subtlety of the effect of words and the images they produce. Only through experimenting with Fuller's substitutions for some period of time can we judge the merits of mathematical terminology that reflects science's new understanding of reality.

Back to the starting point! Nothing can be accepted as self-evident; a new mathematics must be derived though "operational" procedure. Fuller decided that through sufficient observation of both naturally occurring and experimentally derived phenomena without reference to a specific framework, nature's own coordinate system might emerge. He sought a body of generalizations describing the way patterns are organized and able to cohere over time. We shall see how these principles can be discerned both in deliberate experiments with various materials and by recording existing natural patterns.

Bucky's grade-school skepticism was thus the beginning of a lifelong search for "nature's coordinate system." After rejecting traditional academia through his dramatic departure from Harvard's freshman class in 1914,[8] he began an independent exploration of mathematics which he was to pursue for the rest of his life. A sort of philosophical geometry gradually began to take shape, consisting of a rich body of facts and principles, some new and others newly considered. Synergetic geometry is tied together as one cohesive system by the unmistakable presence of Bucky Fuller in conventional and bizarre observations alike. Including expositions called "Tetrahedron Discovers Itself and Universe," "Life," "Cosmic Hierarchy," "Complex of Jitterbugs," and the more conventional "Closest Packing of Spheres," *Synergetics* does not fall neatly into any preconceived category. This volume will attempt to clarify these multifaceted (or polyhedral) observations, which together constitute synergetic geometry.

Nature's Coordinate System

What accounts for the shape similarities among unrelated phenomena, radically different in both scale and material? Or, more fundamentally, what accounts for nature's magnificent orderliness itself? Whether honeycomb or conch shell or virus, time after time individual structures turn out true to form. The fundamental hypothesis behind synergetics—and the work of many other pioneers exploring the science of form—is that nature's structuring occurs according to the requirements of minimum energy, itself a function of the interplay between physical forces and spatial constraints.

Wait. The role of physical forces (gravity, magnetism, electrical and chemical attractions) is clearly important, but what are "spatial constraints"?

We are so used to thinking of "space" as empty nothingness that the idea of its having specific properties seems absurd. However, as will become increasingly clear from the examples throughout this book, *space has shape*. The idea is concisely expressed by Arthur Loeb in his introduction to *Space Structures*: "Space is not a passive vacuum, but has properties that impose powerful constraints on any structure that inhabits it. These constraints are independent of specific interactive forces, hence geometrical in nature."[9] A simple example is the fact that to enclose space with only four polygons, these polygons must all be triangles. Nothing else will work, no matter how hard you try. The limitation is a function of neither material nor size but rather of the nature of space. Fuller alludes to this active role when he says "*natural* is what nature permits."

When Bucky points out that nature doesn't have to stop everything she's doing and gather the physics, chemistry, biology, and mathematics departments to decide how to grow a turnip (or build a virus), he is calling our attention to the self-organization of natural phenomena. Structuring in nature occurs automatically. "Nature has only one department," declares Bucky, "*one comprehensive coordinating system*." How does this self-structuring occur? In the most general terms, according to the path of least resistance, or, as stated above, according to the "requirements of minimum energy." In short, systems automatically find comfortable arrangements, which are necessarily a result of the balance between specific forces and inherent spatial properties. When Fuller set out to inventory possible configurations and thereby formulate generalizations, his exploration was destined to be "geometrical in nature" because of the nature of systems, as we shall learn in Chapter 3. "Nature's coordinate system" is thus a geometry of most economical relationships which govern all structuring. In Fuller's words, "a geometry composed of a system of interrelated vectors may be discovered that represents the complete family of potential forces, proclivities, and proportional morphosis..." (215.02).

We shall see how "operational procedure" produced a geometry of vectors and look at the specific shape of this diagram of potentials in Chapter 7, "Vector Equilibrium," and Chapter 9, "Isotropic Vector Matrix." The most notable characteristic of these models is the absence of perpendicularity. We thus shall explore Fuller's statement that nature is never operating in perpendicular and parallel directions but rather convergently and divergently in radial growth patterns.

Another important aspect of "nature's coordinate system" is the existence of rules governing the coherence of structures. What holds its shape? If nothing is self-evident, we can no longer take "solids," or reliable structures, for granted. Chapter 5 examines Fuller's investigation into structure and the resulting principles governing the stability of systems.

All of this takes a while to sink in. The idea that space has shape is profoundly reorienting; we are so used to conceiving of space as passive emptiness on which we *impose* desired configurations that an entirely new perception cannot be adopted overnight. Nonetheless, upon further study, this premise begins to feel quite comfortable and necessary—an all-embracing *something*ness influencing structural phenomena. As more specifics are uncovered, this conception, which is at once so elusive and so ordinary, begins to seem more and more the latter.

Universe

The ultimate manifestation of nature's coordinate system is "Universe." Fuller deliberately omits the article, for "the universe" implies the possible existence of more than one—just as we do not say "the God" but rather simply "God." Fuller capitalizes "Universe" for the same reason: Universe is everything; it's all there is. (Or, more poetically, "Universe is all that isn't me *and* me."[6]) But Fuller would never leave it at that; he is indefatigably thorough.

Einstein revolutionized our understanding of Universe, explains Fuller; prior to his relativity theory, we could think in terms of a single-frame (simultaneously complete) picture, unimaginably vast, but still simultaneous at any given moment. This understanding must now be replaced by a "scenario" concept:

301.10   Universe is the aggregate of all humanity's consciously apprehended and communicated nonsimultaneous and only partially overlapping experiences.

He willingly disects his own long-winded definition. To Fuller, *aggregate* implies a complex that cannot be comprehended in totality at only one moment (302.00): "*Consciousness* means awareness of otherness." (That one's easy.) To be *apprehended*, information must first be within the range of human perception and then actually be noticed. "*Communicated* means informing self or others. Nonsimultaneous means not occurring at the same time" (302.00); events of

Universe are instead *"partially overlapping"*—like generations. My lifetime overlaps my grandmother's and hers overlaps the life of her grandmother, but I was born long after the death of both my great- and great-great-grandmothers. Such are the events of Universe; every experience overlaps some but not all other experiences.

Another facet of the "scenario" concept centers on the misconception of the environment as a static whole. For instance, looking out at a distant star it is all too easy to think we are "seeing" it just as it is at that moment, while in reality that particular star is so far away that its light takes 100 years to reach us. What we are actually looking at is a "live show" taking place 100 years ago. We are seeing an event that occurred before we were born. Universe is the integral (or sum total) of all experience. It cannot be unitarily conceived, but as thus defined it is all inclusive. "You cannot get out of Universe" (321.02).

Fuller's definition avoids imparting a sense of thingness—part of his effort to encourage us to think in terms of "pure principle." Universe is energy and thought all knotted together by incredibly complicated webs of relationships. It is ultimately impossible to separate the physical and metaphysical; both are "experience." The scientific principles that govern the interactions of energy events—as timeless statements of truth—are themselves metaphysical. The line therefore becomes ever more difficult to draw. This is why Fuller's definition depends upon *consciousness*. Our awareness of energy events defines their existence; we cannot go beyond the limits set by our understanding.

Finally, Fuller assures us that the definition is complete:

People say to me, "I think you have left something out of your definition of Universe." That statement becomes part of my experience. But never will anyone disprove my working hypothesis because it will take experimental proof to satisfy me, and the experiment will always be part of the experience of my definition, ergo included. (306.01)

He elaborates in a 1975 videotaped lecture: "Someone might ask 'what about dreams? I think you left that out,' and I reply, 'no, for that is part of your experience.'" He seems to have all the angles covered.

Generalized Principles

The principle of leverage is a scientific generalization. It makes no difference of what material either the fulcrum or the lever consists.... Nor do the special-case sizes of the lever and fulcrum...in any way alter either the principle or the mathematical regularity of the ratios of physical work advantage....

Mind is the...uniquely human faculty that surveys the ever larger inventory of special-case experiences stored in the brain bank and...from time to time discovers one of the rare scientifically generalizable principles running consistently through all the relevant experience set. (*Synergetics*, p. xxvi)

Fuller spoke frequently and ponderously of the "generalized principles": those statements—be they verbal or in the shorthand of mathematical equations—that have been proven to always hold true. In other words, generalized principles are rules with no exceptions. From the very simple (the mechanical advantage allowed by leverage) to the highly profound ($E = mc^2$, equating matter with energy and quantifying the rate of exchange), these principles, taken all together, describe our Universe. Applying a generalized principle in a novel way is called invention. In the broadest sense, synergetics is the search for generalized principles. ("Design science" is the application, as will be discussed in Chapter 16.)

Fuller placed enormous stock in these principles, and saw humanity's role in Universe as discoverer and utilizer of the progressively uncovered truths. Endowed with "minds" (in contradistinction to "brains," which are merely able to coordinate sensory input), humans are uniquely able to survey successive experiences and detect reliable patterns, thereby discovering over time such subtle workings of Universe as gravity. Utterly invisible and unpredicted by the investigation of separate objects, gravitational force represents a profound discovery and is certainly without exception. More remarkable still, human mind was able to express the magnitude of this force in precise terms: $F = GMm/r^2$.[10] A fantastic leap beyond sensory-based information, this discovery places tiny humans in contact with great Universal motions.

That the human mind is able to detect eternal truths amidst "special-case" experiences, explains Fuller, is the wealth of humanity and our hope for the future. Many of these Universal laws are widely known and applied, as for example the principle of leverage, but other equally reliable truths are all but completely unfamiliar, such as the geometric discoveries described in this book. Fuller, convinced that an inventory of yet to be explored applications had the potential to solve humanity's problems, was the ardent champion of unfamiliar principles.

Universe is thus the total web spun by all generalized principles and their interaction. As rules without exception, they are necessarily "interaccommodative." Fuller's terminology takes some getting used to: "eternally regenerative Universe" is an ongoing event governed by "the omni-interaccommodative complex of unique and eternal generalized principles." The principle of synergy, described in

Chapter 3, accounts for the incalculable complexity of the whole web despite thorough comprehension of many of the separate parts. And finally, "God is the unknowable totality of generalized principles."

Return to Modelability

Synergetics is a product of Fuller's passionate concern with models. Concerned that society's ignorance of science is seriously destructive, he devoted years of thought to ways of alleviating the resulting crisis. In the twentieth century, we suddenly find ourselves confronted with an "invisible" atomic reality in which the average person understands very little about how things work. Although confronted daily with "incredible technology," which to Fuller includes the natural phenomena of Universe as well as the ever-expanding inventory of human invention, the vast majority assume such phenomena to be out of their reach. Fuller attributes this widespread discomfort to both the "invisibility" of science and the devastatingly complicated mathematics without which, scientists claim, their findings cannot be described. The dangerous chasm between scientists and lay people, with the truth guarded by an élite few and the rest resigned to ignorance, thus seems inevitable.

The origin of this troubled state of affairs? An incorrect mathematical system! Long ago human beings surveyed their environment and, seeing a never-ending flat earth, decided upon cubes and orthogonal planes as the appropriate measuring system. Today, says Fuller, we're still stuck with that uninformed early guess, and as a result, nature's behavior has seemed irrational, perverse, and difficult to explain because we're using the wrong kind of yardstick. With accurate models, he claims, this ignorance can be eradicated. The purpose of synergetics is to make the invisible events and transformations of Universe visible, through tangible models that elucidate the *principles* behind our energy-event Universe. Human beings will thereby be able to "coordinate their senses" with a new understanding of reality.

Synergetics is full of tantalizing models; the difficulty comes in assigning them to aspects of physical reality. However, a number of notable examples, in which a newly discovered scientific phenomenon is described by one of Fuller's previously developed models, suggest that there may be many more such successes to come. The immediate goal therefore is to unravel and study the geometric system itself.

2

The Irrationality of Pi

Young Bucky Fuller seems to have been haunted by π. Discovering that the ratio of a sphere's circumference to its diameter is an irrational number commonly referred to by the Greek letter π was perhaps the most disturbing of all geometry's strange lessons. Just as he could not imagine where its accomplice in deception, the infinite straight line, stopped, Bucky could not let go of a vision of thousands upon thousands of digits trailing after 3.14159..., spilling out the classroom window and stretching absurdly to the next town and then farther, never allowed to stop.

He has told the story countless times, each rendition sounding like the first, with an air of revelation. If he happens to be standing by the ocean, so much the better; he's bound to talk about the foaming waves.

"Look at them all"! he marvels, "beautiful, beautiful bubbles, every one of them!" Bucky recalls that growing up near the ocean gave him plenty of time to think about the structuring of these bubbles. Looking back at the wake trailing behind his boat, or standing knee-deep in breaking waves at the shore, he saw the water continuously being laced with white foam. Its whiteness was created by vast numbers of tiny air bubbles, each one suddenly formed and emerging at the water's surface. "How many bubbles am I looking at"? Bucky would ask himself, "fantastic numbers, of course." He could not help wondering: if each and every one of those spheres involves π, to how many digits does nature carry out the irrational π in making one of those bubbles, before discovering that it can't be completed? At what point does nature stop and make a "fake bubble"? And how would the decisions be made? In meetings of the chemistry and mathematics departments? No, the young Bucky concluded, I don't think nature's making any fake bubbles; I don't think nature is using π.

It is a decidedly amusing image: nature gathering department heads, nature's consternation over fudging the numbers, getting away with imperfect bubbles—and it is through such deliberate personification that the ideas become memorable. The story of a young man looking out from his ship carries a deeper message. Fuller uses this particular event—even assigning a time and place, 1917 on a U.S. Navy ship—as a moment of revelation, the threshold of his conscious search for nature's coordinate system. The scene is bait, to draw us in, make us as curious about nature's structuring as that young sailor was. It works. If he had started on a heady discourse about mathematical concepts versus natural structures, his audience might have walked away.

It's important to realize the nature of his rebellion: not to challenge the theoretical numerical ratio between the circumference and diameter of the ideal sphere, but rather to challenge that sphere to materialize. Irrational numbers don't belong in tangible experiences. It is a question of sorting out the demonstrable from the impossible and then developing models based on the former. Essentially, Bucky is choosing not to play with π, posing the question "why shouldn't mathematics deal with experience"?

"Nature Isn't Using Pi"

Nature can have no perfect spheres because she has no continuous surfaces. The mathematician's sphere calls for all points on its surface to be exactly equidistant from the center. This "sphere," explains Fuller, has no holes. It is an absolutely impermeable container sealing off a section of Universe, a perpetual energy-conserving machine defying all laws of nature. The illusion of a physical continuum in any spherical system is due to the limitations of the human senses.

On some level of resolution, all physical "solids" and surfaces break down into discrete particles. A magnifying glass uncovers the tiny dots of different colors that make up the "pure" blue sky in a magazine photograph, and a microscope—if it could be set up in the sea—would reveal that nature's bubbles are likewise fragmented, consisting of untold numbers of discrete molecules located approximately equidistant from an approximate center. If such a phenomenon could be precisely measured, we would find that the ratio of a bubble's circumference and diameter is some number very close to the elusive π, but the point, says Bucky, is that the bubble differs from the Greek ideal. A physical entity is necessarily demonstrable and finite, while an irrational number such as π is just the opposite. A *real* sphere consists of a large but finite number of interconnected individual energy events. Moreover, the middle of each of the implied chordal (straight-line) connections between events is slightly closer to the sphere's center than the ends, thereby violating the mathematical definition, and insuring some small departure from the ratio π. Any cross-section of this sphere will be, not a circle, but a many-many-sided polygon. Likewise, the most perfect-looking circle, carefully drawn with a sharpened pencil and accurate compass, will appear fuzzy and fragmented through a good magnifying glass. The most precise ball bearing is imperfect: a good approximation of π, but always imperfect, with bizarre mountains and valleys dramatically visible after 5000 times magnification. In Chapter 15 we shall examine the specifics of one alternative model for the mathematician's impossible sphere.

"Fake bubbles" led to further contemplation about π. Nature is always associating in simple whole rational numbers, thought Bucky: H_2O, never $H_\pi O$. Irrational numbers do not show up in chemical combinations of atoms and molecules. Clearly, nature employs "one coordinate, omnirational, mensuration system." (410.011)

None of this is new information; any one of us could have thought about it for a while and understood that of course π does not play a role in the making of bubbles. What happens is that turbulence introduces air into the water and the air is so light that it floats toward the surface to escape. It's an easy problem of getting the most air in each pocket with the least surface area of water pushing in to collapse the bubble. A spherical space allows the most volume per unit of surface area (as we shall learn in later chapters), providing the most efficient enclosure for that air. (Nature is exquisitely efficient.) But, even if we did contemplate this mystery at the seashore, we probably didn't bother to take our children aside and explain it to them before they became perplexed and frustrated by the never-ending π in mathematics class.

If nature's lack of employment of π isn't news, why does Bucky persist? Why use his own considerable—but still finite—energy to lecture tirelessly on the subject for half a century? The answer lies in the fact that our education and popular awareness fall short of the mark; Fuller felt that society did not sufficiently emphasize this discrepancy between theoretical games and real structures. We introduce "solids" to school children long before they learn the real story about energy-event reality. If humanity is to feel comfortable with science, argues Fuller, education must present accurate models in the first place. The message behind the bubble's story is: why not tell the truth from the beginning? He calls for widespread recognition of *discrete energy events*, a reality that cannot be perceived by human senses.

Unfortunately, the issue is not that easily resolved. We shall not ever completely escape the brain-teasing presence of irrational numbers. For example, the construction of the simplest polygons, with carefully measured unit-length edges, frequently produces irrational diagonals. The distance between nonadjacent vertices in a regular pentagon is the irrational number known as the "golden section," which we shall see again in Chapter 11, and the diagonal of a square is incontrovertibly $\sqrt{2}$. However, in every actual construction, a finite approximation can be determined according to the specific limitations of available measuring instruments. That numerical fractions must terminate is a consequence of investigating verifiable experience rather than theoretical relationships. Just as architects do not work with irrational-length two-by-fours, measurements obtained through "operational procedure" are necessarily real. Finally, Bucky's dismissal of these troublesome values calls attention to the granular constitution of physical reality, as described in the following section.

Finite Accounting System

"You cannot have a fraction of an energy event," philosophizes Bucky. Science's progressive subdivision ultimately reaches indivisible particles, which means that reality consists of whole numbers[1] of energy events. Therefore, we need models that will demonstrate the concept of structures consisting of discrete, or countable, units. In synergetics, area and volume are presented as quantities that can be counted, as opposed to measured and described as a continuum. Area is associated with some whole number of events on the surface, while volume is tallied as the number of events throughout a system. Chapter 8 will examine how sphere-packing models illustrate the association of volume and surface area with discrete units; Chapter 10 will discuss the concept of volume in general and its role in synergetics.

We are not used to thinking of reality as submitting to a "finite accounting system," but in fact it does. Consider the phenomenon of light. Few things seem more continuous than the light that fills your living room when you turn on the switch, or the light that seems to fill the whole world on a sunny day at the beach. But we now know that even light consists of individual packages of energy, called photons. Although they could hardly be smaller or lighter, photons are nonetheless discrete events, countable in theory, even if there are always too many, jumping around too fast, to make that enumeration possible in practice. With its flawless illusion of continuity contradicting science's detection of constituent photons, light is a perfect phenomenon to symbolize the validity of a "finite accounting system." We could hypothesize a system in which volume was tallied in terms of the number of photons in a given space, and area, the number of photons found at the surface; however, it would be a fantastically impractical system. A more appropriate unit is called for. For now, however, we simply consider the qualitative implications of the discrete-events concept. A punctuated reality is hard to get used to.

Which Way Is "Up"?

Fuller proposed a revolution in modes of thinking and problem-solving, which above all else required a comprehensive approach, as will be discussed in Chapter 16. To Fuller, "comprehensive" means not leaving out anything—least of all humanity's important tool of language. A dictionary contains an inventory of 250,000 agreements, he explains, specific sounds developed as symbols for 250,000 nuances of experience. He saw this gradual accomplishment as one of the most remarkable developments in the history of humanity, with its implied cooperative effort.

One aspect of his revolution thus involves an effort to employ words accurately. Fuller's discourses on the subject tend to be quite humorous, almost (but not quite) concealing how deeply serious he was about the matter. Much of our language is absolutely stuck in "dark ages" thinking, he would lecture. *Up* and *down*, for instance. These two words are remnants of humanity's early perception of a flat earth; "there is no up and down in Universe!" exclaims Fuller. When we say "look down at the ground" or "I'm going downstairs" we reinforce an underlying sensory perception of a platform world. Neither the ground nor Australia can accurately be referred to as down; three hours after a man in California says that the astronauts are up in the sky, the shuttle is located in the direction of his feet. Up and down are simply not very precise on a spherical planet. The replacements? *In* and *out*. The radially organized systems of Universe have two basic directions: *in* toward the center and radially *out* in a plurality of directions. Airplanes go out to leave and back in to land on the earth's surface. We go in toward the center of the earth when we walk downstairs. The substitutions seem somewhat trivial at first, but again it is difficult to judge without trying them out. Experimenting with "in" and "out" can be truly reorienting; unexpectedly one does feel more like a part of a finite spherical system—an astronaut on "Spaceship Earth." (It's hard to keep at it for long however; up/down reflexes are powerful.)

The sun does *not* go down, insists Bucky; how long are we going to keep lying to our children? First of all, we now know there is no up and down in the solar system, and secondly, the sun is not actively touring around the earth. Rather, we are the travelers, and our language should reflect that knowledge. Oddly, the phraseology which gives the sun an active role ("darling, look at the beautiful sun going down") does seem to reinforce the erroneous conception of a yellow circle traveling across the sky. While we *know* this isn't the case, it often *feels* like the way things happen. This effect is not easily measured and probably varies from individual to individual.

What does all this have to do with geometry? Remember that one of the goals of synergetics is to help coordinate our senses with reality, that is, to put us in touch with Universe, which to Fuller involves eradication of the erroneous vocabulary which keeps us locked into "dark-ages" thinking on a sensorial level. In short, we need to align our reflexes with our intellect. Instead of sunset and sunrise, reinforcing the sun's active role, Fuller suggests *sunclipse* and *sunsight*, which imply instead that our view of the sun has been obscured and that an obstacle has been removed, respectively. Who knows if people could adjust to such substitutions? It might be worth a try.

The cube is another remnant of flat-earth days, and Fuller has a wealth of reasons to prefer another mathematical starting point, which will be discussed throughout this volume.[2] Our age-old dependence on squares and cubes is honored by an unfortunate verbal shorthand for "x to the second power" and "x to the third power." The expressions "x squared" and "x cubed" are so commonly used that most people assume these multiplication functions to have a true and exclusive relationship to squares and cubes. The shorthand "squared" is derived of course from the fact that a square can be subdivided by parallel lines, with "x" subdivisions along each edge, into "x to the second power" smaller squares. For example, a square with 2 "modular subdivisions" per edge contains 4 small squares, and similarly 3 subdivisions yield 9 squares, 4 yield 16, 5 yield 25, and so on (Fig. 2-1). Everyone is familiar with these diagrams, but what most people do not realize is that this result is not unique to squares. Triangles exhibit the same property, as also shown in Figure 2-1. Furthermore, explains Fuller, triangles take up only half the

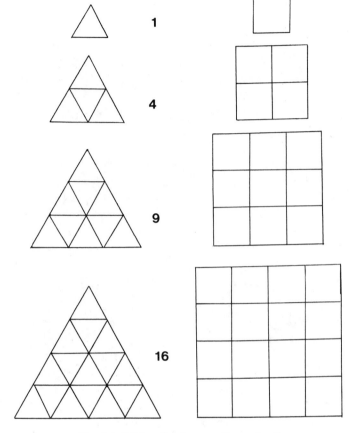

Fig. 2-1. "Triangling" versus "squaring."

space, because every square divides into two triangles. Triangles in general, therefore, provide a more efficient diagram for the mathematical function of multiplying a number by itself. Nature is always most economical; therefore nature is not "squaring"; she is "triangling."

Happily, the same is true in three dimensions: the increasing volumes of subdivided tetrahedra (if this is an unfamiliar shape, wait until the next chapter!) supply the third-power values just as accurately as do subdivided cubes: $2^3 = 8$, $3^3 = 27$, $4^3 = 64$, etc.[3] Tetrahedra of course take up less room than cubes, and to Bucky, the choice is clear. We do not, at this point in the text, have the necessary experience to fully understand the third-power model shown in Figure 2-2, but the relevant principles will eventually be discussed. The point to be made now is that squares and cubes cannot boast a special inherent significance for multiplicative

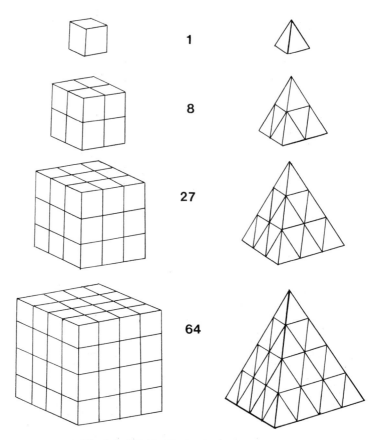

Fig. 2-2. "Cubing" versus "tetrahedroning."

accounting. Triangles and tetrahedra are equally reliable (and in some ways more reliable, as we shall see). Fuller argues that our arbitrary habitual references to squares and cubes keep us locked into a right-angled viewpoint, which obscures our vision of the truth. From now on, says Bucky, we have to say "triangling" not "squaring" if we want to play the game the way nature plays it.

Visual Literacy

Fuller's concern with fine-tuning communication, developing and using words that are consistent with scientific reality, is one facet of the role of language with respect to synergetics. Another deals with the difficulty of describing visual and structural patterns. Anyone who has tried to describe an object over the telephone is well aware of the problems involved; there seems to be a shortage of functional words. The temptation to use your hands is irresistible, despite the futility. This scarcity of linguistic aids is especially severe for non-cubical structures—which characterize most of nature. That a language of pattern and structure is not widely accessible indicates that humanity's understanding of such phenomena is similarly underdeveloped. We thus join forces with Fuller in an investigation of this neglected field, and in so doing we become more and more aware of the rich complexity of the order inherent in space. Along the way, we are introduced to some new terminology that includes the lesser-known language of geometry as well as some words invented by Fuller. More information—useful only, it is clear, if the terminology is both precise and consistent—leads to better comprehension, which in turn leads to the ability to experiment knowledgeably. Experiment fosters both greater understanding and invention—in short, progress.

In conclusion, Fuller's insistence on employing accurate vocabulary is part of an important aspect of human communication. We join him as pioneers in the science of spatial complexity, the terminology of which is for the most part unfamiliar. The systematic study of structural phenomena is an important and badly neglected aspect of human experience.

Peaceful Coexistence

There is a strong temptation to ignore synergetics on the grounds that we feel perfectly able to handle mathematical concepts that cannot be seen. Academic "sophistication" leaves us with a certain intellectual pride that makes Fuller's observations with their child-like (but-the-emperor-isn't-wearing-any-clothes) ring to them seem unimportant. Every child is boggled by infinity and surfaces of no thickness, but these are necessary concepts, natural extensions of philosophical "what-ifs." The human mind is not bounded by the constraints of demonstrability.

True enough. However, it is also possible to define a system of thought and exploration that is confined to the "facts of experience," and moreover such a system is able to reveal additional insights about physical and metaphysical phenomena that would not necessarily be discovered following the traditional route. Such is the case with Fuller's synergetics; as we shall see, his hands-on approach led to a number of impressive geometrical discoveries. Synergetics is a different kind of mathematical pursuit, not a replacement for calculus but rather a complementary body of thought. Reading further will not face you with an ultimate demand for a decision of fundamental allegiance: synergetics *or* the mathematics you learned in school. Rather, you have the option of being additionally enriched by a fascinating exploration of structure and pattern that cannot help but change the way you see the visual environment. When Bucky reminisces, "There is nothing in my life that equals the sense of ecstasy I have felt in discovering nature's beautiful agreement," he offers us an enticing invitation. He has taken an alternate route and it has not disappointed him.

Operational mathematics cannot claim exclusive rights to the name of mathematics, any more than other branches of mathematics can; it is simply a new approach, stemming from the characteristically human drive to experiment. Its claim instead is that exposure to these concepts fosters an understanding of nature's structuring and therefore provides an advantageous base of experience. In short, synergetics is an internally consistent system which has produced significant models with respect to certain physical phenomena and led directly to practical inventions (with "life-support advantage," to use Fuller's terminology). It is likely that thus far we have only skimmed the surface of the applications of synergetics; very few people have been exposed to its principles, and so their full significance is as yet untested.

Finally, Bucky's approach is compellingly playful, and we should read his material with the inclination to enjoy the adventure. "Sense of ecstasy"? Why not? Fuller's unorthodox way of looking at mathematics—and indeed Universe—can provide a way to circumvent some of our more rigidly held assumptions. Here is an invitation to start over; with a temporary suspension of disbelief, we can embrace a new understanding of the exquisitely designed "scenario Universe." One of the challenges of synergetics lies in opening rusty mental gates that block discovery, for we are asked to be explorers and "comprehensive thinkers"—job titles not usually assigned in our specialized world. The essence of synergetics is "modelability"; anyone can play with these models, and likewise, claims Fuller, anyone can understand science once they get their hands on nature's coordinate system. Its accessibility and emphasis on experimental involvement makes Fuller's thinking extremely important. He offers us an approach to learning and thinking—an open-minded, experimental curiosity—which itself is applicable to every discipline and aspect of life. No question is too simple or too complex to be asked. Fuller's first words in *Synergetics* are "Dare to be naïve," reminding us that we have the option to see the world through new eyes.

3

Systems and Synergy

Fuller's mathematical explorations seem to fly out in many directions at once, but they share a common starting point in the concept of systems. Derived from the Greek for "putting together," the word *system* means any group of interrelated elements involved in a collective entity. If that sounds vague, it's meant to. The theme is widely encompassing.

Long ago, secluded in his room experimenting with toothpicks or ping-pong balls or whatever available material seemed likely to reveal nature's secrets, Fuller began to see a persistent message of interdependence. He was later to discover the precisely descriptive word "synergy," but even without that lexical advantage a sense of interacting parts increasingly dominated his vision. More like the poets and artists of his generation than the scientists, he was drawn to relationships rather than objects.

By stating that Fuller looked at systems, we learn very little, especially in view of the word's current popularity. We have transportation systems and systems analysts, stereo systems and even skin-care systems, all conspiring to diminish the precision and usefulness of the word. But let us enter into the spirit of Bucky's half-century search and abandon our twentieth-century sophistication in order to rediscover the obvious en route to the surprising and complex. Much can be gained, for alongside our era's growing consciousness of systems and interdependence is also its ever-increasing specialization. Individuals are encouraged to narrow their focus, precluding a comprehensive vision and inhibiting curiosity. How-things-work questions are reserved for children; as adults we are afraid to step outside our expertise. Furthermore, we are quite likely to have decided that we have no use for mathematics at all by the time we reach high school. Both factors—specialization and avoidance of mathematics—cause some aspects of Fuller's synergetics to seem dense while others seem oddly simple. However, the novelty of his approach serves to give us new insights, so bear with

Bucky as he "discovers the world by himself";[1] his enthusiasm may be contagious.

Fuller was unafraid to appear naïve. He announced his observations with equal fervor for the simplest ("only the triangle holds its shape") and the very complex (the surface angles of the planar rhombic triacontahedron correspond exactly to the central angles of the icosahedron's fifteen great circles), but on every level he was conscious of systems.

A system, says Bucky, is a "conceivable entity" dividing Universe into two parts: the inside and the outside of the system. That's it (except, of course, for the part of Universe doing the dividing; he demands precision). A system is anything that has "insideness and outsideness." Is this notion too simple to deserve our further attention? In fact, as is typical of Fuller's experimental procedure, this is where the fun starts. We begin with a statement almost absurdly general, and ask what must necessarily follow. At this point in Fuller's lectures the mathematics sneaks in, but in his books the subject is apt to make a less subtle entrance! (Half-page sentences sprinkled with polysyllabic words of his own invention have discouraged many a reader.) The math does not have to be intimidating; it's simply a more precise analysis of our definition of system.

So far a system must have an inside and an outside. That sounds easy; he means something we can point to. But is that trivial after all? Let's look at the mathematical words: what are the basic elements necessary for insideness and outsideness, i.e., the minimum requirements for existence?

Assuming we can imagine an element that doesn't itself have any substance (the Greeks' dimensionless "point"), let's begin with two of them. There now exists a region between the two points—albeit quite an unmanageable region as it lacks any other boundaries. The same is true for three points, creating a triangular "betweenness," no matter how the three are arranged (so long as they are not in a straight line). In mathematics, any three noncollinear points define a plane; they also define a unique circle.

Suddenly with the introduction of a fourth point, we have an entirely new situation. We can put that fourth point anywhere we choose, except in the same plane as the first three, and we invariably divide space into two sections: that which is inside the four-point system and that which is outside. Unwittingly, we have created the minimum system. (Similarly, mathematics requires exactly four noncoplanar points to define a sphere.) Any material can demonstrate

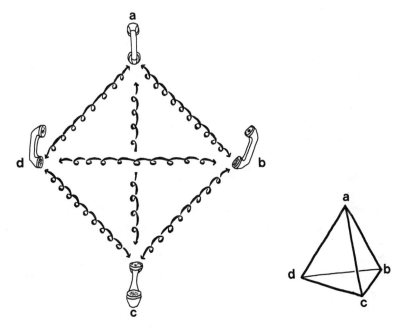

Fig. 3-1. Six connections between four events, defining a tetrahedral system.

this procedure—small marshmallows and toothpicks will do the trick, or pipecleaner segments inserted into plastic straws. The mathematical statement is unaltered by our choice: a minimum of four corners is required for existence.

What else must be true? Let's look at the connections between the four corners. Between two points there is only one link; add a third for a total of three links, inevitably forming a triangle (see if you can make something else!). Now, bring in a fourth point and count the number of interconnections. By joining *a* to *b*, *b* to *c*, *c* to *d*, *d* to *a*, *a* to *c*, and finally *b* to *d* (Fig. 3-1), we exhaust all the possibilities with six connections, or *edges* in geometrical terminology. Edges join *vertices*, and together they generate windows called *faces*.

This minimum system was given the name *tetrahedron* (four sides) by the Greeks, after the four triangular faces created by the set of four vertices and their six edges (Fig. 3-1). Fuller deplored the Greek nomenclature, which refers exclusively to the number of faces—the very elements that don't exist. ("There are no solids, no continuous surfaces... only energy event complexes [and] relationships."[2]) However, he did not fully develop a satisfactory alternative, so we shall have to work with the time-honored convention. What we lose in accurate description of physical reality, we gain in clarity and consistency.

The tetrahedron shows up frequently in this exploration. This and other recurring patterns seem coincidental or magical at first, but soon come to be anticipated—endless demonstrations of the order inherent in space. The process is typical of synergetics: we stumble into the tetrahedron by asking the most elementary question—what is the simplest way to enclose space?—and later, everywhere we look, there it is again, an inescapable consequence of a spectrum of geometric procedures.

The straightforward logic of our first encounter with the tetrahedron drives us to wonder if it displays any other unique properties. It turns out to be a reliable sort of minimum module or "quantum," as Fuller points out in myriad ways. Not the least impressive involves counting the edges of all regular, semiregular, and triangulated geodesic polyhedra (from the simple cube to the more complex rhombic dodecahedron to the vast array of geodesics.) The resulting numbers are all multiples of the tetrahedron's six. We can therefore take apart any polyhedral system in these categories and reassemble its edges into some number of complete tetrahedra. Even though we are not yet familiar with these other polyhedra, the observation stands as a representative example of the surprising whole-number relationships which make our investigation increasingly alluring.

Conceptual and Real Systems

Notice that these geometrical systems are purely conceptual: so far they exist only in our mind, as sets of relationships. They can be lent substance by any number of materials, as for example the toothpicks and marshmallows mentioned above. However, the essence of a system is independent of the choice of materials: six sticks will create a tetrahedron whether we use wood or metal. Similarly, four Ping Pong balls or four people constitute a tetrahedral system. The tetrahedron, being a conceptual entity, is "sizeless and timeless." Thus Fuller writes in *Synergetics*, "Size is always a special-case experience." (515.14) Size belongs to a different category of parameters than vertices, edges and faces—those which only relate to actual constructs, such as color, temperature, and duration.

Does he take this concern with terminology too far? His justification is twofold, encompassing first a deep conviction that words influence the shape of our thinking, and secondly faith in the power of accurate models in problem-solving. For humanity to solve its complex problems, he was convinced that vocabulary and other models had to be absolutely precise. So Fuller's concern that we recognize conceptual systems as sizeless sets of relationships capable of being physically embodied is an essential part of his geometry of thinking. In either form, the emphasis is on the relationships.

We begin to see a basis for the phenomenon of vastly different properties exhibited by systems with identical constituents. One notable example is the soft grey graphite of pencils in contrast to sparkling impenetrable diamonds, both consisting exclusively of carbon atoms. Geometry alone accounts for their differences. We shall see how later, but as always our attention will be on shape and valency (numbers of connections) rather than substance.

Let's get back to our starting point. Any subdivision of Universe constitutes a system. We have found the simplest example and learned the mathematical terms; our next step is to look at more complex systems. It does not stretch our definition to discuss some very elaborate forms, such as a school, or even a crocodile (both have the requisite boundary). For that matter, our entire planet is a system—unimaginably intricate but still finite. This line of thought, together with our geometry lesson, suggests an approach to problem-solving: a "whole systems" view that demands consideration of the influence of every move on its entire system. Such an approach, which prohibits short-term or piecemeal solutions to long-term problems, may sound simplistic or vague at first; however, the method is based on the assumption of rigorous analytical procedures.

In *Synergetics*, Fuller introduces the concept with a deliberately simple example, which provides an analogy for more complex situations. A fictitious child draws on the ground with a stick, announcing that he has made a triangle. Then Bucky himself intervenes to point out to the child that he has created *four* triangles—not just one—because "operational mathematics" requires that a triangle must be inscribed *on something* in order to exist. Whether on a piece of paper or on the surface of the earth, that something is always a system, with an inside and an outside. Unwittingly, the child has divided the earth's surface into two areas. Both regions are bounded by three arcs, and therefore both qualify as *spherical triangles*,[3] despite the fact that one is small and tangible and the other covers most of the earth's surface. We are not used to thinking in these terms, philosophizes Fuller, but we must begin to really think about what we're doing.

Hold on, says the child, that's only two triangles! Why did you say there were four? Well, Bucky continues, concave and convex[3] are not the same; when you delineated two concave triangles on the outside surface, you also created two convex spherical triangles—one very small and the other very large—on the inside. But I didn't *mean* to make four triangles, protests the bewildered child. That doesn't matter, his teacher replies; you are still responsible for them.

His story can be considered a parable; its purpose is as much to encourage a sort of holistic morality as to make a mathematical statement. The message: tunnel vision is obsolete. As human beings, we cannot afford to ignore the effect of our actions on the rest of a system while working on an isolated part. Rather we must become responsible for whole systems. We didn't *mean* to make four triangles—or indeed, to make "the big mess of pollution" (814.01).

As playful as this example seems, it calls our attention to what Fuller perceives as a dangerous "bias on one side of the line" inherent in traditional mathematics. He points out that our grade-school geometry lessons involve concepts defined as *bounded* by certain lines—a triangle is an area bounded by three lines, for example—thus excusing us from paying attention to its environment. Once a figure is delineated, we no longer have to consider the rest of the system. This narrow approach, Fuller argues, instills in us at an impressionable age a deep bias toward *our* side of the line; we see and feel an unshakable correctness about our side's way of "carrying on." On the other hand, "Operational geometry invalidates all bias" (811.04). It forces us to remain aware of all sides. In Fuller's opinion, being taught in the first place that all four triangles are "equally valid" would significantly influence our later thinking and planning.

One consequence of this approach is Fuller's realization that "unity is inherently plural" (400.08). "Oneness" is impossible, he explains, for any identifiable system divides Universe into two parts, and requires a minimum of six relationships to do so. Furthermore, as illustrated in the above parable, all "operations" produce a plurality of experiences, and awareness itself—without which there can be no life—implies the existence of "otherness." Ergo, "Unity is plural and at minimum two."[4]

Limits of Resolution as Part of the Whole-Systems Approach

Another important aspect of Fuller's systems concept is *tune-in-ability*, which deals with limits of resolution and is best explained by analogy. Fuller's ready example, as implied by the term, would be to remind us of the radio waves of all different amplitude and frequency, filling the room wherever you happen to be reading this page. These waves are as much a part of physical reality as the chair you are sitting in, but the specific energy pattern is such that you cannot tune in to the programs without help from a radio. Information and energy are scattered chaotically throughout your room, mostly undetected, except for the small fraction (chairs, visible light, and so on) that can be directly perceived by human senses. You can turn on the radio and thereby tune in to one program (one system), temporarily ignoring the rest.

Boundaries change all the time as new elements are incorporated into a system, or as the focus zooms in to investigate a component in greater detail. New levels of complexity reveal distinct new systems. For example, we might look at the system called your living room, and then want to consider its function in the bigger system, your house, or conversely, zoom in to investigate the red overstuffed chair, and the details of its carpentry and upholstery; or further still, one nail might be of interest as a system. We can also go back out—to your town, your state, etc. The concept of tune-in-ability allows us to treat a set of events or items as a system despite the involvement of many concurrent factors on other levels.

What kinds of things constitute systems? Tetrahedron, crocodile, room, chair, you, thought, ... Wait.

What about thoughts? We recall Fuller's lifelong effort never to use mankind's precious tool of language carelessly: "I discipline myself to define every word I use; else I must give it up." In a 1975 videotaped lecture, he explains that he would not allow himself the use of any word for which he did not have "a clear experientially referenced definition."[4] Such an effort requires enormous discipline to avoid automatic associations and thereby enable an objective analysis of each word. It extends to the most basic words and actions —even "thinking." Fuller formulates his definition analytically, asking, "What is it I am conscious of doing, when I say I am thinking"? We may not be able to say what it *is*, but we should be able to specify the procedure.

Thinking, he explains, starts with "spontaneous preoccupation"; the process is never deliberate initially. We then *choose* to "accommodate the trend," through conscious dismissal of "irrelevancies" which are temporarily held off to the side, as they do not seem to belong in the current thought. Fuller places "irrelevancies" in two categories: experiences too large or too infrequent to influence the tuned-in thought, and those too small and too frequent to play a part. The process he describes is similar to tuning a radio, with its progressive dismissal of irrelevant (other-frequency) events, ultimately leaving only the few experiences which are "lucidly relevant," and thus interconnected by their relationships.

Thinking isolates events; "understanding" then interconnects them. "Understanding is structure," Fuller declares, for it means establishing the relationships between events.

A "thought" is then a "relevant set," or a "considerable set": experiences related to each other in some way. All the rest of experience is outside the set—not tuned in. A thought therefore defines an insideness and an outsideness; it is a "conceptual subdivision of Universe." "I'll call it a system," declares Bucky; "I now have a geometric description of a thought."

This is the conclusion that initially led Fuller to wonder how many "events" were necessary to create insideness and outsideness. Realizing that a thought required at least enough "somethings" to define an isolated system, it seemed vitally important to know the minimum number—the terminal condition. He thereby arrived at the tetrahedron. "This gave me great power of definition," he recalls, both in terms of understanding more about "thinking" and by isolating the theoretical minimum case, with its four events and six relationships.

One example of the development of a thought—by no means a minimal thought—could be found in what to cook for dinner. Walking to the grocery store, you notice that the leaves of the maple trees are turning autumn-red, but you consciously push that observation off to the side to be considered later, as it does not relate to the pressing issue of dinner. You begin to pull in the various relevant items: the food that you already have at home that could become a part of this meal, what you had for dinner last night, special items that might be featured by the grocery store, favorite foods, how they look, ideas about nutrition, certain foods that go well together, and so on. Out of this jumble of related events, a structure starts to take form. After a while, dinner is planned, and your mind is free to attend to some of the other thoughts waiting quietly in the side chambers.

This kind of digression is typical of Fuller's discourse, both written and oral. Such juxtapositions of geometry and philosophy are quite deliberate, for synergetics strives to identify structural similarities among phenomena—both physical and metaphysical. Fuller encourages us to seek these patterns, which we often miss because of the narrow focus of our attention.

To conclude: Geometry is the science of systems—which are themselves defined by relationships. (Geometry is therefore the study of relationships; this makes it sound relevant to quite a lot!) A system is necessarily polyhedral; as a finite aggregate of interrelated events, it has all the qualifications. Relationships can be polyhedrally diagrammed in an effort to understand the behavior of a given whole system. Along these same lines, Fuller has described synergetics as the "exploratory strategy of starting with the whole and the known behavior of some of its parts and the progressive discovery of the integral unknowns along with the progressive comprehension of the hierarchy of generalized principles" (152.00). This mouthful can readily be identified as Fuller's elaboration of Eddington's definition of science as "the systematic attempt to set in order the facts of experience."[5]

Thinking in terms of systems is a crucial part of Fuller's mathematics. The isolation of systems enables the description of local processes and relationships without reference to an absolute origin—an indispensable tool in a scenario universe. And finally, we pay particular attention to how Fuller's geometry emerges—its principles developing from the basis of the process of thinking. Hence the title of Fuller's opus: *Synergetics: The Geometry of Thinking.*

Synergy

Implicit in the above discussion of systems is a property described accurately by only one word. "Synergy" has come into fairly widespread use recently, perhaps due to Bucky's many years of championing its cause, or perhaps just because we finally needed it badly enough. Formerly unknown except to biologists and chemists, this word describes the extraordinarily important property that "the whole is more than the sum of its parts." In Fuller's words, "Synergy means the behavior of whole systems unpredicted by the behavior of their parts taken separately."

Consider the phenomenon of gravity. The most thorough examination of any object (from pebbles to planets) by itself will not predict the surprising behavior of the attractive force between two objects, in direct proportion to the product of their masses and changing inversely with the square of the distance between them. Another dramatic example is the combination of an explosive metal and a poisonous gas to produce a harmless white powder that we sprinkle on our food—sodium chloride, or table salt.

Bucky's favorite illustration was the behavior of alloys: "synergy alone explains metals increasing their strength" (109.01). He enthusiastically describes the properties of chrome-nickel steel, whose extraordinary strength at high temperatures enabled the development of the jet engine. Its primary constituents—iron, chromium, and nickel—have tensile strengths of 60,000, 70,000, and 80,000 pounds per square inch respectively, and combine to create an alloy with 350,000 psi tensile strength. Not only does the chain far exceed the strength of its weakest link, but counter-intuitively even outperforms the *sum* of its components' tensile capabilities. Thus the chain analogy falls through, calling for a new methodology which will incorporate interaugmentation.

A flood of examples of "synergy" is so readily available that one might wonder how we got along without the word. Bucky wondered and concluded that humanity must be out of touch with its environment. Synergy is certainly how nature works; though we pay little direct attention to the phenomenon, we are still familiar with it. Few are surprised by complex systems arising out of the interaction of simpler parts.

Fuller took it a step further. He saw the age-old forms of geometry as models of synergy, comprehensible only in terms of relationships. His eye drawn to their vector edges, he simply did not perceive the "solid" polyhedra of Plato. Self-exiled from the formal mathematical community which would have told him otherwise, Fuller saw the static constructs of geometry as ready and waiting to elucidate the dynamic events of physical Universe.

Determined to model the new "invisible" energetic reality, Fuller began to refer to his accumulated findings as "energetic geometry." As the search for "nature's coordinate system" progressed and the recurring theme of synergy became more and more prevalent, the term evolved to "synergetic-energetic geometry" and finally to "synergetics." Fuller's vocabulary tended to develop organically in response to his changing needs for emphasis. He felt a great responsibility to get it just right. In the thirties, enchanted by certain properties of the cuboctahedron, Fuller replaced the Greek name with his own trademark word, "Dymaxion," less from egotism than from frustration in being unable to invent exactly the right name—one with enough impact.[6] Later, he found the perfect term to express its unique property, and eagerly renamed this indispensible shape "vector equilibrium." The name remained unaltered; when Fuller found his truth, he never wavered.

The term "synergetics," then, was a response to the single most important characteristic of energetic reality. As discussed in Chapter 1, Fuller's overriding goal was to collect the "generalized principles."

The law of synergy, although too all-encompassing to seem a valid starting point for such an inventory, dictates a basic strategy of starting with a whole system and then investigating its parts. The most painstaking study of its separate components will never reveal the behavior of a system. All other generalized principles therefore must be subsets of this fundamental truth: the whole is not equal to the sum of its isolated parts.

Now we take some time out to look at aspects of conventional geometry that will illuminate Fuller's work, despite the fact that it is not directly included in *Synergetics*.

4

Tools of the Trade

Whether or not the thought of high-school geometry class stirs unpleasant memories, chances are that most of the actual material is long forgotten. Moreover, few of geometry's more pleasing properties are taught, leaving volumes of elegant transformations in the realm of esoteric knowledge. Lack of exposure to these age-old discoveries is the primary barrier to understanding synergetics.

By now familiar with Fuller's underlying assumptions, we shall take time out to introduce some background material. The origins of humanity's fascination with geometry can be traced back four thousand years, to the Babylonian and Egyptian civilizations; two millennia later, geometry flourished in ancient Greece, and its development continues today. Yet most of us know almost nothing about the accumulated findings of this long search. Familiarity with some of these geometric shapes and transformations will ease the rest of the journey into the intricacies of synergetics.

A little experimentation with basic geometric forms and procedures reveals the important role of space itself. The work of other thinkers reinforces the fundamental premise of synergetics: *space has shape*—all structures are formed according to spatial symmetries and constraints. It turns out that the number of symmetrical arrangements allowed by space is surprisingly limited; perpetual (synergetic) interaction of relatively few patterns accounts for the seemingly endless variety of form.

Even for readers whose background in geometry is already strong, reviewing it can be enjoyable and perhaps even illuminating. There are so many significant connections among polyhedral shapes and operations that even experienced geometers continually discover new ones.

Bucky's delight in a new-found truth never lost its intensity. He promptly adopted each discovery into his growing synergetics inventory. If, in his enthusiasm, he appeared to be taking credit for age-old discoveries, let us—rather than judging—try to enter into the spirit of his search. And if egotism seemed to have gone hand in hand with enthusiasm, it is because both grew out of his constant willingness to see everything as if for the first time. Both were part of the whole system Bucky.

Our most rewarding course is to immerse ourselves in the geometry. We shall first set the stage with some of the Greek basics, and then move on to the work of Arthur L. Loeb of Harvard University —whose "Contribution to Synergetics"[1] provides an analytical counterpart to the 800 pages of Fuller's more intuitive approach—before moving on to the thought-provoking twists of Fuller's mathematical thinking.

As this thinking is encompassing and holistic, it would be counter-productive to scrutinize isolated parts of Fuller's geometry. Synergetics must be critically examined as a whole system before judging its contribution to mathematics. The emphasis (and certainly the excitement) in Fuller's ongoing research was in the inherent "omni-inter-accommodation"—that is, uncovered principles are never contradictory but rather augment each others' significance. Fuller's developing inventory was thus continually strengthened, becoming more and more integral to his thinking. He was especially gratified, in his later years, to receive numerous letters from noted scientists, reinforcing his discoveries with their observations from research in other fields.

Bucky's guiding purpose in developing synergetics was to reacquaint us with Universe. Such eagerness is compelling. ("Beautiful, beautiful bubbles . . . ," "eternally regenerative Universe," " . . . ecstasy in discovering nature's beautiful agreement.") We do him and ourselves a disservice to expect his approach to fit comfortably into the traditional scholarly framework. Unacknowledged by academic institutions for most of his life, he felt unencumbered by their conventions. (But we are not, and this volume will attempt to locate sources wherever possible.)

Onward! Let's take another look at geometry; the journey itself is fascinating, as well as full of useful tools for understanding Fuller's work.

Plato's Discovery

The Platonic polyhedra, as their name suggests, have been around for almost two millenia. However, despite the fact that the five shapes (also called regular polyhedra) are well known, few people are aware of what exactly defines this group—and fewer still of the implications about space itself.

The requirements seem lenient at first. They are two: the faces of a polyhedron must be identical, and the same number of them must meet at each vertex. The tetrahedron fits, with its four triangles and four equivalent (three-valent) vertices.[2] (Notice the roots of the word equivalent.) From these two criteria alone, one might suspect the existence of many more regular polyhedra.

Triangles

Let's study the possibilities step by step, beginning with the simplest polygon (fewest sides) and the smallest number of edges meeting at each vertex. In keeping with Fuller's use of vectors as edges, this study will be confined to "straight" edges. It is quickly apparent that a minimum of three edges must meet at each vertex of a polyhedron, for if vertices join only two straight edges, the resulting array is necessarily planar. This lower limit can also be expressed in terms of faces, for we can readily visualize that a corner needs at least three polygons in order to hold water—which is another way of saying the inside is separated from the outside, as specified by Fuller's definition of system.

The minimal polygon is a triangle. Three triangles around one vertex form a pyramid, the base of which automatically creates a fourth triangular face. As all corners and all faces are identical, the first regular polyhedron—a tetrahedron—is completed after one step (Fig. 4-1a).

Next, a second regular polyhedron can be started by surrounding one vertex with four triangles, resulting in the traditional square-

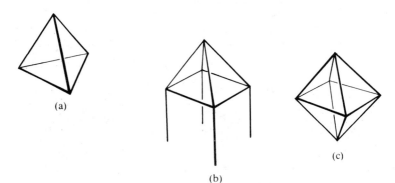

(a)

(b)

(c)

Fig. 4-1

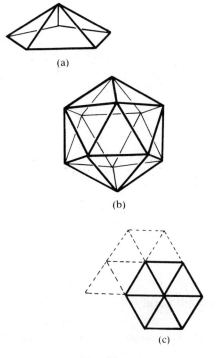

(a)

(b)

(c)

Fig. 4-2

based pyramid (Fig. 4-1b). But, as our specifications for regularity indicate that all vertices must connect four edges, an additional edge is required at each of the four vertices around the pyramid's base, bringing the total to twelve edges. By connecting the four dangling edges (Fig. 4-1c), we introduce a sixth vertex, which is also surrounded by four triangles. The result is an octahedron, with eight triangular faces and six four-valent vertices. Both criteria for regular polyhedra are satisfied, and so the octahedron is added to our list.

As the procedure is thus far simple and successful, analogy suggests the next step. Five triangles around one vertex form a shallow pyramid (Fig. 4-2a). Paying attention as before to nothing but the two rules, we continue to employ triangular faces while making sure that five of them surround each corner. The structure essentially builds itself in that there is only one possible outcome—and we don't even have to know what it is to be able to finish the task. The icosahedron, with twenty (in Greek, "icosa") triangles and twelve five-valent corners, is indeed regular (Fig. 4-2b).

Once more then. Bring six triangles together at a corner. But wait! Six 60-degree angles add up to 360 degrees, or the whole plane (Fig.

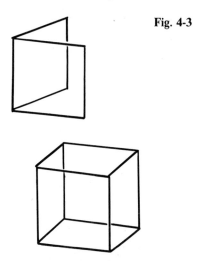

Fig. 4-3

4-2c). An unprecedented result, this indicates that we could surround vertices with six equilateral triangles indefinitely and never force the collection to curve around to close itself off. A space-enclosing system is therefore unattainable with exclusively six-valent vertices. Is it possible we have exhausted the possibilities for regular polyhedra out of triangles? We simply cannot have fewer than three or more than five triangles around all vertices and create a closed finite system. We thus encounter a first upper limit.

Squares

Let's back up and start the procedure over, with another kind of polygon. Continuing the step-by-step approach, we change from triangles to squares by increasing the number of sides by one. The method is reliable if plodding. What happens if two squares come together? Again, that's just a hinge (Fig. 4-3). So start with three. If there are three squares around one corner, the same must be true for all corners, and as before, the structure is self-determining. Continue to join three squares at available vertices until the system closes itself off. With six squares and eight corners, this is the most familiar shape in our developing family of regular polyhedra—the ubiquitous cube.

Next, we gather four squares at a vertex and immediately hit ground. Four times 90 degrees is 360 degrees, the whole plane again. And while such groups of four will generate graph paper indefinitely, they can never close off as a system. So that's it for squares.

Fig. 4-4

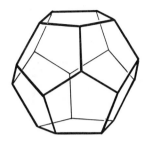

Pentagons

Triangles, squares,..., now we come to pentagons. Three of them around each corner works. A system exists as soon as there are twelve pentagons and twenty three-valent corners. Called the *pentagonal dodecahedron* ("dodeca" is Greek for twelve), it adheres to the definition of regular polyhedra (Fig. 4-4).

Try four around a corner. Another new situation. The interior angle of a regular pentagon measures 108 degrees (see Appendix A); four of them together add up to 432 degrees, which is more than the planar 360. This indicates that four pentagons simply will not fit around one point. Not only have we reached a stopping point for regular polyhedra out of pentagons, but this example also shows that not all regular polygons can be made to fill a page (or tile a floor). Specific spatial constraints apply in two dimensions as well as in three. We shall investigate these patterns in more detail in Chapter 12.

Having exhausted the pentagonal possibilities, we go on to hexagons. Three of their 120-degree angles total 360 degrees, and are therefore planar right away—creating the hexagonal pattern seen in honeycomb and frequently used for bathroom floor tiles (Fig. 4-5). Three heptagons with angles totalling 385.71 degrees just won't fit together. Neither will octagons, nor any polygons with more vertices.

A Limited Family

A quick review reveals the surprisingly limited inventory of five regular polyhedra: tetrahedron, octahedron, icosahedron, cube, and pentagonal dodecahedron (Fig. 4-6). Like it or not, we have reached the end. A child in kindergarten, with the two rules carefully explained, will discover the same five shapes. Space takes over, imposing that upper limit. This idea runs counter to the bias of our mathematical background that space is passive emptiness and we

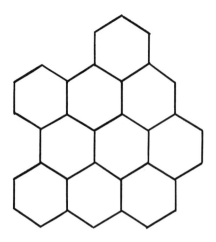

Fig. 4-5

impose desired configurations. Invisible, unyielding constraints sound more like mysticism than science.

If we dwell on this subject, that is because it is crucial to understanding the framework of Bucky's investigation. Space has specific characteristics, and we want not only to list and understand them, but also to begin to really *feel* their embracing qualities—a sense of structured space permeating all experience.

The regular polyhedra provide a good starting point from which to branch out in all directions. An eighteenth-century mathematician, Leonhard Euler (1707–1783), greatly simplified our task with his

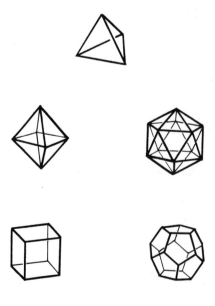

Fig. 4-6. Five regular polyhedra.

realization that all patterns can be broken down into three elements: crossings, lines, and open areas. He thereby introduced the basic elements of structure (vertices, edges, and faces) which underlie all geometrical analysis. Bucky saw this contribution as a breakthrough of equal importance to the law for which Euler is known, for this precise identification of terms enabled Euler's other, more famous observation.

Euler's Law

Euler's law states that the number of vertices plus the number of faces in every system (remember Fuller's definition) will always equal the number of edges plus two. It may not sound like much at first, until you reflect on the variety of structures—from the tetrahedron to the aforementioned crocodile—that all obey this simple statement. Every system shares this fundamental relationship. The number of vertices can be precisely determined by knowing the number of faces and edges, and so on.

Denote the numbers of vertices, faces, and edges by V, F, and E respectively. Then we have $V + F = E + 2$. What about that constant 2? The other numbers might be extremely large, or as small as four, yet by Euler's equation the difference between the number of edges and the sum of the number of vertices and faces will always be exactly two. It seems unlikely at first.

To gain confidence in this principle, let's try it out on the regular polyhedra. Remember the four vertices and four faces of the tetrahedron; four plus four is eight, exactly two more than its six edges. Not bad so far. Similarly, we can count to check the other four regular polyhedra. The results are displayed in Table I.

The persistent 2 has led to some controversy. Fuller long ago assigned his own meaning to the recurring number: 2 occurs in the

Table I

	V	F	E	2	Total
Tetrahedron	4 +	4 =	6 +	2	8
Octahedron	6 +	8 =	12 +	2	14
Cube	8 +	6 =	12 +	2	14
Icosahedron	12 +	20 =	30 +	2	32
Pentagonal dodecahedron	20 +	12 =	30 +	2	32

equation to represent the "poles of spinnability." That requires some clarification.

All systems, Fuller explains, can be spun about a central axis. An axis has two poles (e.g., north and south), and thus at any given time, two vertices must be poles. Subtract the two poles from the total number of vertices to get a number exactly equal to the combined number of edges and faces. Simply stated, in Fuller's view, the permanent 2 represents two vertices which, as "poles of spinnability," should be subtracted from the total number of vertices to equalize the equation. It's a puzzling explanation.

There is another way to view that constant element, as you may have guessed. What are the characteristics of the variables included in Euler's Law? Zero-dimensional points, one-dimensional lines, and two-dimensional areas—each a level higher than the last. The law compares all aspects of structure—almost. Something's missing. Three-dimensional space is the next and only absent parameter; its geometrical units (corresponding to vertices, edges, and faces) are cells. Why are cells left out of this fundamental relationship? The other view, as expounded by Loeb, says they're not.[3]

Recall Bucky's definition of a system: a subdivision of space creating an inside and an outside, both equally important. Two cells! Could the constant 2 in the equation be incorporating the otherwise only missing dimension? Indeed it is. Euler's law is actually a special case of Schlaefli's formula for any number of cells. In other words, if the number of cells, C, is substituted for 2, the equation holds true for multicellular structures, that is, arrangements with more than two cells: $V + F = E + C$, even when C is greater than 2.[4]

Evaluation of significance is a tricky business, but we cannot avoid indulging in it altogether, as Fuller's *Synergetics* overflows with such speculation. It is ultimately puzzling that Fuller, with his emphatic observation that every polyhedron is a system dividing Universe into two parts (inside and outside), would not connect Euler's constant 2 with the implied two cells. His insistence that both parts of a system must be considered equally important provides a truly new orientation in geometry.

At some point in any discussion about Euler and the "polar two," Fuller would speak of a structural system's inherent "constant relative abundance." The meaning of the term eluded many. Fuller observes in *Synergetics* that the number of faces in triangulated systems is always two times the number of vertices minus two (the subtraction, he says, again taking account of the two poles).[5] He further states that the number of edges is *three* times the number of vertices less two.

Turning these two statements into simple equations, we have

$$F = 2(V - 2)$$

and

$$E = 3(V - 2).$$

Simplifying,

$$F = 2V - 4, \qquad E = 3V - 6,$$
$$F + 4 = 2V, \qquad E + 6 = 3V.$$

Combining the two equations by subtraction, we have

$$\begin{array}{r} E + 6 = 3V \\ -(F + 4 = 2V) \\ \hline E - F + 2 = V, \end{array}$$

or

$$E + 2 = V + F.$$

So his observations directly substantiate Euler's law. Constant relative abundance refers to the everpresent two faces and three edges for each vertex in triangulated polyhedra, (excepting of course the two "poles," which are not included, according to Fuller's rationale).

Duality

Table I reveals a curious pattern. Notice the relationship between cube and octahedron, along with the similar pairing of pentagonal dodecahedron and icosahedron. Polyhedra thus related, each with the same number of vertices as the other has faces, are called each other's *dual*.[6] We are thereby introduced to duality as a numerical relationship; the vertex and face tallies are simply switched. Reassuringly, the significance extends: Loeb observes the geometric manifestation of duality in precise matching of vertex to face. Two dual polyhedra line up with every corner of each meeting the center of a window of the other, as the correspondence implies. (Figure 4-7a shows the dual relationship of the cube and octahedron.)

We have conspicuously ignored one member of Plato's polyhedral family. What is the tetrahedron's dual? Following Loeb's example, we count elements to predict the answer, and find the same number of vertices as faces. Therefore, by interchanging the two elements to find the tetrahedron's dual, we generate another tetrahedron (Fig.

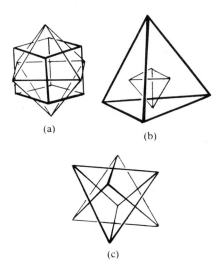

(a)

(b)

(c)

Fig. 4-7. (a), (b) Dual polyhedra. (c) Two tetrahedra intersect to form the eight vertices of a cube: "star tetrahedron."

4-7b). Once again, the minimum system stands out: only the tetrahedron is its own dual. Put two of them together to check: the four windows and corners line up, and curiously, the combined eight vertices of two same-size tetrahedra outline the corners of a cube, as Fuller never tired of explaining. ["Two equal tetrahedra (positive and negative) joined at their common centers define the cube" (462.00, figure; Fig. 4-6c).]

Truncation and Stellation

The introduction of two operations explored by Arthur Loeb will further elucidate the relationships between polyhedra. By altering and combining the five regular polyhedra, we can generate new shapes—one of the geometric phenomena that inspired Fuller's coinage "intertransformabilities." Loeb's observations will help to clarify the meaning of this polysyllabic Fullerism.

"Truncation" involves chopping off corners so that they are replaced by surfaces (Fig. 4-8a).[7] Truncation of a three-valent vertex will generate a triangle; four-valent vertices become squares, and so on. The number of edges determines the number of sides of the new polygon. Notice that the definition does not specify how much of the corner is sliced away in truncation. Loeb's work reinforces our

(a)   (b)   (c)   (d)

Fig. 4-8. Degenerate truncation of a tetrahedron.

emphasis thus far on topology (studying numbers of elements, or valency) rather than size. The location of slicing is therefore unimportant, until the truncation planes move inward far enough to touch each other. At that point the edges between the new planes disappear, and so the topology changes. Figure 4-8 shows various possibilities, including that final chop at the mid-edge point. This special limit case, called by Loeb *degenerate*, yields some interesting results, as we shall see. (For example, the "degenerate truncation" of a tetrahedron unexpectedly turns out to be another member of the Platonic Family, the octahedron, as revealed by Fig. 4-8d.)

Readers interested in learning more about these concepts and the mathematical analyses involved should read *Space Structures* by Arthur Loeb. Although we introduce only "vertex truncation," Loeb's studies extend to "edge truncation" as well. For our purpose of becoming familiar with the interconnectedness of basic polyhedra, we explore just one of Loeb's discoveries in the following pages. Then when we see these shapes again in the context of synergetics, some of their important relationships can be anticipated.

Whereas truncation cuts off corners, "stellation" is accomplished by the *addition* of a corner—imposing a shallow pyramid on a formerly flat face. How many sides the pyramid will have is de-

Fig. 4-9. Stellated cube and octahedron.

881321

termined by the number of edges of the face to be stellated. Figure
4-9 compares a stellated cube and a stellated octahedron. Notice that
—as required by their different faces—the cube's superimposed
pyramids have four sides, while the octahedron's have three. *Degenerate stellation* occurs when the altitude (height) of the added pyramids is such that adjacent faces of neighboring pyramids become
coplanar, as will be illustrated below.

An Experiment

What happens if we truncate both members of a pair of dual
polyhedra? The octahedron and cube provide a representative pair
for Loeb's elegant experiment. Chop off the corners of the cube,
creating eight new triangle faces, while changing the six squares into
octagons (Fig. 4-10a). The truncated octahedron (also called tetra-

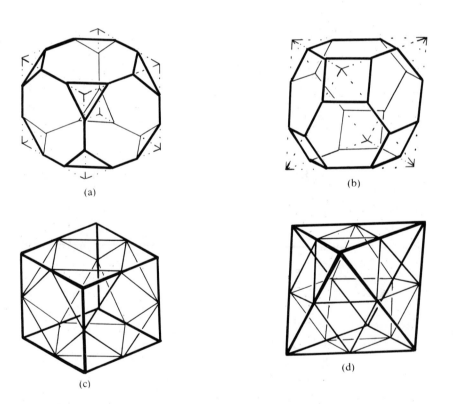

(a)

(b)

(c)

(d)

Fig. 4-10. (a), (b) Truncation of cube and octahedron. (c), (d) Degenerate truncation
of cube produces the same polyhedron as degenerate truncation of octahedron:
cuboctahedron.

kaidecahedron for its fourteen faces) (Fig. 4-10b) is a space-filling shape—a subject that we shall explore more fully in Chapter 12.

Now, notice what happens if we allow all truncation planes to expand. The truncated cube's triangles and octahedron's squares independently spread to the "degenerate case," with truncation planes meeting at mid-edge points (Fig. 4-10c, d). Suddenly octagons and hexagons are phased out, becoming squares and triangles respectively. The two different systems turn into the same polyhedron. Conventionally called the *cuboctahedron* for reasons now apparent (and renamed "vector equilibrium" by Fuller, for a reason explored in Chapter 7), this polyhedron plays a crucial role in *Synergetics*. In the present context it merely provides an exemplary illustration of the interactions of duality and truncation.

Loeb's further investigation reveals that degenerate truncation of *any* dual pair leads to the same polyhedron. For instance, degenerate truncation of both the pentagonal dodecahedron and the icosahedron creates the *icosadodecahedron*, named for its twenty triangles and twelve pentagons. The pentagons, created by slicing the twelve five-valent vertices, alternate with the twenty icosahedral triangles. Looked at another way, twenty triangles have resulted from chopping off all the three-valent vertices of the pentagonal dodecahedron (Fig. 4-11). With either outlook, we begin to see the effect of duality. (Take special note of the system's twelve fivefold elements; their presence will soon be interpreted as a fundamental law under certain conditions.)

We now need a new category as we uncover new polyhedra that are not regular, but certainly far from random or irregular. The cuboctahedron and the icosadodecahedron are alike in having only one kind of vertex but two different kinds of faces. Such polyhedra are called *semiregular*. It will not come as a surprise that their duals have one kind of face and two kinds of vertices. You can begin to

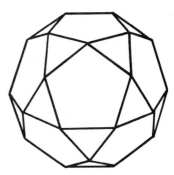

Fig. 4-11. Icosadodecahedron.

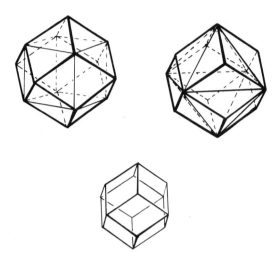

Fig. 4-12. Degenerate stellation of cube produces the same polyhedron as degenerate stellation of octahedron: rhombic dodecahedron.

imagine the potential for generating new systems, as reflected in the term "intertransformabilities."

The elegant results of degenerately truncating dual polyhedra inspire further questions. Does it follow that degenerate *stellation* of a dual pair will create the same polyhedron? And if so, how do we define degenerate stellation? Experimentation answers both questions.

To check the hypothesis with our reliable octahedron–cube pair, we stellate both systems independently. Try to imagine the transition: six shallow square pyramids are superimposed on the cube, while eight triangular pyramids are added to the octahedron, as seen in Figure 4-9. We then increase the altitude of all pyramids, until triangles of adjacent pyramids just become coplanar. In both cases, twenty-four individual triangular facets suddenly merge into twelve rhombic (or diamond) shapes, thereby creating the *rhombic dodecahedron*, named for its twelve rhombic faces. The original edges of the cube form the short diagonals of the twelve faces, while the octahedral edges turn into the twelve long diagonals (Fig. 4-12). We thereby have illustrated degenerate stellation.

Notice that both members of the original dual pair have the same number of edges. This turns out to be a necessary condition of duality, which follows logically from the nature of the geometric correspondence. In order for each vertex to line up with a face, the edges of two dual polyhedra must cross each other, as can be seen in

Figure 4-6. We further observe that whereas degenerate *truncation* produced quadri*valent* vertices (cuboctahedron), degenerate *stellation* produced quadri*lateral* faces (rhombic dodecahedron). A generalized principle emerges.

"Intertransformability"

As Fuller so often observed, nature consists exclusively of endlessly transforming energy. Atoms gather in "high-frequency" clusters, disassociate periodically, and regroup elsewhere—new patterns, different substances. We interpret these transient events as solids and liquids because of the limitations of our five senses. To Fuller, "exquisitely transformable" polyhedra were highly logical models with which to elucidate nature's behavior. "Intertransformability" thus applies to both nature and her models.

We now consider a final experiment from Loeb's research, to bring the subject to a close. Start by truncating a representative polyhedron such as the cube. Continue to the degenerate polyhedron (in this case the cuboctahedron, as illustrated in Fig. 4-10c), and put it aside for a moment. Now take that same cube, the degenerate stellation of which yields the rhombic dodecahedron (as seen in Fig. 4-12) with its twelve rhombic faces.

Twelve identical faces? Have we accidentally overlooked a regular polyhedron? No, for the other requirement, of identical vertices, is not satisfied. The diamonds' obtuse angles[8] come together at eight three-valent vertices, and their acute angles[8] at six four-valent vertices. (Refer to Fig. 4-12.)

Those numbers are familiar. Twelve identical faces, along with six four-valent and eight three-valent vertices, correspond to the twelve identical vertices, six four-valent faces (squares), and eight three-valent faces (triangles) of the cuboctahedron (Fig. 4-13). Loeb thus

Fig. 4-13. Cuboctahedron and rhombic dodecahedron are dual polyhedra.

shows that duality extends to semiregular polyhedra. This fact would have enabled us to predict the rhombic dodecahedron's existence, by specifying twelve similar quadrilateral faces to correspond to twelve four-valent vertices, and so on. Now we can examine the results of the experiment. A pair of dual polyhedra was created by applying the two inverse operations to the same initial shape.

Had we started with an octahedron, the cube's dual, the results would have been the same. Degenerate stellation creates the rhombic dodecahedron (Fig. 4-8), and degenerate truncation, the cubocta-hedron, as described above (Fig. 4-9). Thus, the two operations generate dual polyhedra from one starting point. Loeb concludes that degenerate truncation and stellation are *dual operations*.

To generalize, Loeb discovers that if both members of a pair of dual polyhedra are truncated (or both stellated) to the degenerate case, they will lead to the same result. Conversely, separate degener-ate truncation and stellation of the same shape create a new dual pair. Finally, regular polyhedra usually generate semiregular ones.

Symmetry

The foundation is almost in place. A final tool to pick up is an understanding of symmetry: "exact correspondence of form or con-stituent configuration on opposite sides of a dividing line or plane or

M

S

(a)

R

(b)   **Fig. 4-14**

about a center or axis." This somewhat abstruse definition from *The American Heritage Dictionary* introduces the two types of symmetry.

Mirror symmetry is the more familiar, involving the exact reflection of a pattern on either side of a "mirror line" (or plane). The letter "M" exhibits mirror symmetry; "R" does not (Fig. 4-14b).

Rotational symmetry specifies that a configuration can be rotated some fraction of 360 degrees (depending on the numerical type of rotational symmetry) without changing the pattern. For example, a square, exhibiting fourfold rotational symmetry about its center, can be rotated 90, 180, or 270 degrees without detectable change. Similarly, an icosahedron has fivefold rotational symmetry about an axis through a pair of opposite vertices (Fig. 4-14a). The letter "S" exhibits twofold rotational symmetry; it looks the same after a 180-degree turn (Fig. 4-14b). In other words, in x-fold rotational symmetry, constituents of a pattern are repeated x times about a common center.

These concepts will prove useful as we proceed through Fuller's discoveries.

5

Structure and "Pattern Integrity"

"I'm a little child and I've just found my mother's necklace."

Famous for his marathon lecture sessions, Fuller used to talk about synergetics for days on end. Time constraints in later years usually prohibited such extensive coverage, but he almost always told the story of the necklace. Looking a bit like a small child himself as he draped this ten-foot loop over his shoulders, Bucky explained that the process of collecting "experimental evidence" starts with children.

Fuller was a remarkable teacher, particularly in his ability to explain difficult concepts in simple terms. Not drawn to the formal logic of proofs, his genius lay in his novel use of everyday experiences. Elaborately detailed descriptions, relying on familiar materials and specific colors, were tailored to elucidate various complex phenomena. In one such scenario he is able to explain the intimidating concept of precession, which is one of the mysteries of gyroscopic motion, through a series of easily visualized events.[1] His images materialize so vividly in the mind's eye that the underlying abstract statements can be grasped effortlessly.

"I'm going to be a little child now." We are immediately in his world, looking out. Even having heard this routine countless times, one can forgive the simplicity of the story. This is stuff for five-year-olds, but it is riveting—and a welcome break in the often heady lecture.

The necklace grows out of Fuller's insistence that every child is born a genius—endlessly curious, probing, full of wonder about everything. If a child's questions are rewarded with answers that feel right, that is, correspond to his experience, the inherent genius will blossom. More often, not challenged creatively by tedious memorization that doesn't seem to relate to the world around him, a child simply learns to play the game. Fuller's conviction that children spontaneously leap at the chance to understand Universe when excited by true and comprehensive information was a primary motivating force behind synergetics. His aim was to supply models to elucidate the wonders of science to adults and children alike.

It's an unusual necklace. Ten or more thin wooden dowels are linked together with red rubber-tubing segments into a continuous flexible loop. Bucky keeps taking it off his shoulders to remove another one of the ten-inch dowels. One by one they drop to the floor, as the necklace turns into recognizable polygons. Recognizable, that is, when he struggles to hold them out flat and round. Soon, the "drapable necklace" resembles a hexagon, then a pentagon. Next, the four sticks that are easily persuaded to be a square, just as readily collapse into a bundle—four parallel sticks held in one fist (Fig. 5-1).

His expression is utterly earnest, "You remember when the teacher went to the blackboard and drew a square?" (Nods fill the lecture hall.) "Well, the only reason it stayed a square was that the blackboard held it there!" The shape collapses, dangling from one hand. (Many laugh. Some look concerned; they have begun to sense that he is deeply serious about this.)

One more stick is pulled out of the loop. Three are left dangling. If he removed another, the necklace would disappear, for two sticks alone cannot form an open loop. Connect the two ends of the three-dowel string, and suddenly, "*It holds its shape*," he cries out, astonished. Loudness underscores the importance of this fundamental truth, with enthusiasm undiminished by the repetition of a thousand lectures. "Only the triangle is inherently stable" (609.01).

Bucky reminds us that the conditions and materials of the experiment did not change. That red tubing is still flexible, the sticks still

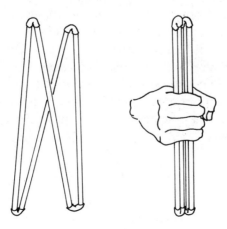

Fig. 5-1

Fig. 5-2. Leverage.

rigid. So what is responsible for the sudden change? Before going on, he wants us to really understand *why* a triangle holds its shape.

Two sticks connected by a hinge create two lever arms. The farther out a force is applied, the greater the mechanical advantage—which means that forces of decreasing strength can accomplish the same result (Fig. 5-2). Each flexible corner is stabilized with minimum mechanical effort by a force exerted at the very ends of its two sticks, or lever arms. A third stick, acting as a "push–pull brace," can be attached to the ends of the other two sticks, to most efficiently stabilize the flexible opposite angle. So each of the triangle's three sticks "stabilizes its opposite angle with minimum effort." Only a triangle has a built-in bracing device for each corner; therefore, only a triangle is stable.

Pattern Integrity

The term "pattern integrity" is a product of Fuller's lifelong commitment to vocabulary suitable for describing Scenario Universe. He explains,

When we speak of pattern integrities, we refer to generalized patterns of conceptuality gleaned sensorially from a plurality of special-case pattern experiences.... In a comprehensive view of nature, the physical world is seen as a patterning of patternings.... (505.01–4)

Let's start with his own simplest illustration. Tie a knot in a piece of nylon rope. An "overhand knot," as the simplest possible knot, is a good starting point. Hold both ends of the rope and make a loop by crossing one end over the other, tracing a full circle (360 degrees). Then pick up the end that lies underneath, and go in through the opening to link a second loop with the first (another 360-degree turn). The procedure applies a set of instructions to a piece of material, and a pattern thereby becomes visible.

What if we had applied the same instructions to a segment of manila rope instead? Or a shoelace? Or even a piece of cooked spaghetti? We would still create an overhand knot. The procedure does not need to specify material. "A pattern has an integrity independent of the medium by virtue of which you have received the information that it exists" (505.201). The knot isn't that little bundle that we can see and touch, it's a weightless design, made visible by the rope.

The overhand-knot pattern has integrity: once tied, it stays put. In contrast, consider directions that specify going around once (360 degrees), simply making a loop. This pattern quickly disappears with the slightest provocation; it is not a pattern integrity. (Even though the overhand knot depends on friction to maintain its existence, a single loop will not be a stable pattern no matter how smooth or coarse the rope.) Notice that it requires a minimum of two full circles to create a pattern integrity. $2 \times 360 = 720$ degrees, the same as the sum of the surface angles of the tetrahedron (four triangles yield 4×180 degrees). Minimum system, minimum knot, 720 degrees. A curious coincidence? Synergetics is full of such coincidences.

A similar example involves dropping a stone into a tank of water. "The stone does not penetrate the water molecules," Fuller explains in *Synergetics*, but rather "jostles the molecules," which in turn "jostle their neighboring molecules" and so on. The scattered jostling, appearing chaotic in any one spot, produces a precisely organized cumulative reaction: perfect waves emanating in concentric circles.

Identical waves would be produced by dropping a stone in a tank full of milk or kerosene (or any liquid of similar viscosity). A wave is not liquid; it is an event, reliably predicted by initial conditions. The water will not surprise us and suddenly break out into triangular craters. As the liquid's molecular array is rearranged by an outside disturbance, all-embracing space permeates the experience. Because liquids are by definition almost incompressible, they cannot react to an applied force by contracting and expanding; rather, the water must move around. In short, the impact of any force is quickly distributed, creating the specific pattern shaped by the interaction of space's inherent constraints with the characteristics of liquid.

The concept thus introduced, Bucky goes on to the most important and misunderstood of all pattern integrities: life. "What is really important...about you or me is the *thinkable you* or the *thinkable me*, the abstract metaphysical you or me,...what communications we have made with one another" (801.23). Every human being is a unique pattern integrity, temporarily given shape by flesh, as is the knot by rope.

> ...All you see is a little of my pink face and hands and my shoes and clothing, and you can't see *me*, which is entirely the thinking, abstract, metaphysical me. It becomes shocking to think that we recognize one another only as the touchable, nonthinking biological organism and its clothed ensemble. (801.23)

Our bodies are physical, but life is metaphysical. Housed in a temporary arrangement of energy as cells, life is a pattern integrity far more complex than the knot or the wave. Remember that all the material present in the cells of your body seven years ago has been completely replaced today, somehow showing up with the same arrangement, color, and function. It doesn't matter whether you ate bananas or tuna fish for lunch. A human being processes thousands of tons of food, air, and water in a lifetime. Just as a slip knot tied in a segment of cotton rope, which is spliced to a piece of nylon rope, in turn spliced to manila rope, then to Dacron rope (and so on) can be slid along the rope from material to material without changing its "pattern integrity," we too slide along the diverse strands supplied by Universe—as "self-rebuilding, beautifully designed pattern integrities." No weight is lost at the moment of death. Whatever "life" is, it's not physical.

The key is consciousness. "Mozart will always be there to any who hears his music." Likewise, "when we say 'atom' or think 'atom' we are...with livingly thinkable Democritus who first conceived and named the invisible phenomenon 'atom'" (801.23). Life is made of awareness and thought, not flesh and blood. Each human being embodies a unique pattern integrity, evolving with every experience and thought. The total pattern of an individual's life is inconceivably complex and ultimately eternal. No human being could ever completely describe such a pattern, as he can the overhand knot; that capability is relegated to the "Greater Intellectual Integrity of Eternally Regenerative Universe."[2]

If we seem to stray from the subject of mathematics, resist the temptation to categorize rigidly. Synergetics does not stop with geometry. Fuller was deeply impressed by a definition in a 1951 Massachusetts Institute of Technology catalog, which read "Mathematics is the science of structure and pattern in general" (606.01): not games with numbers and equations, but the tools for systematic analysis of reality. To Fuller this meant that mathematics ought to enable the "comprehensivist" to see the underlying similarities between superficially disparate phenomena, which might be missed by the specialist. Rope may not be much like water, but the knot is like the wave—is like the tetrahedron.

Our emphasis thus far has been on pattern. What about structure?

Let's go back to the regular polyhedra. On constructing the five shapes out of wooden dowels and rubber connectors, it is immediately apparent that some are stable and others collapse. The "necklace" demonstrates that only triangles hold their shape, and so the problem becomes quite simple.

Picture a cube. Better yet, make one out of dowels and rubber tubing, or straws and string. Whatever material you choose, as long as the joints are flexible, the cube will collapse. Connectors with a certain degree of stiffness, such as marshmallows or pipecleaners, are misleading at first, because the cube appears to stand on its own. However the shape is so easily rearranged by a slight push that the illusion does not last.

The six unstable windows must be braced in order for a cube to be rigid. So six extra struts, inserted diagonally across each face, would be the minimum number that could stabilize the system.

Six struts? Just like a tetrahedron! And not only are there the right number of struts, but they can also be arranged the same way. A regular tetrahedron fits inside a cube with its six edges precisely aligned across the cube's faces (Fig. 5-3). We can therefore state that there is an implied tetrahedron in every stable cube. Nothing in our investigation thus far would predict the precise fit of a cube and a tetrahedron. This and many other examples of shared symmetry among polyhedra (as seen in the previous chapter) are powerful demonstrations of the order inherent in space.

If the cube is unstable without a scaffold of triangulation, what about cardboard models? They stand up by themselves with no trouble. The key word is cardboard. A polyhedron constructed out

Fig. 5-3. Inscribed tetrahedron stabilizes cube.

of stiff polygon faces, rather than edges and connectors, is effectively triangulated. Cardboard provides the necessary diagonal brace. It provides a lot of extra material too, but the untrained eye will not recognize the redundancy at first.

Likewise, the stiffness of marshmallows or pipecleaners provides triangulation, in the form of tiny web-like triangular gussets at the corners, strong enough to stabilize the whole window. Furthermore, stiff material is itself rigid because of triangulation on the molecular level.

Only when polyhedra are considered as vector systems is stability an issue. Construction is one method of determining stability, but a simple formula utilized by Loeb can also be used to check: $E = 3V - 6$.[3] If the number of edges is less than three times the number of vertices minus six, the system will be unstable. But the criterion is really even simpler: for polyhedra without interior edges, a stable system is always triangulated and a triangulated system is always stable. $3V - 6 = E$ holds true for a polyhedral shell if and only if that system consists exclusively of triangles.

The other option is to establish internal triangles; an unstable shell can be stabilized by interior edges, or body diagonals. Instead of triangulating the surface, the bracing members create triangles inside the shell to maintain the system's stability.

Structure

Upstaged by the crowd of oversized polyhedral toys, Bucky again resembles the small child we saw earlier, playing with his mother's necklace. But the words this time are more ambiguous: "There are only three basic structural systems in Universe."

Fuller long ago decided that science simply did not have a definition of structure, and took it upon himself to remedy the oversight. Science, Fuller explains, did not feel the need for a definition, because "structure" seemed to be self-evident. It holds its shape! Language, caught in the old world of "solids," did not keep up with science's evolving understanding of the true nature of matter. In a universe consisting entirely of fast-moving energy, we must ask how something holds its shape.

"Structure is defined as a locally regenerative pattern integrity of Universe" (606.01). A good starting point. Structure is also "a complex of events interacting to form a stable pattern." Similar, but more specific: the pattern consists of action, not things.

"Regenerative" is an important qualification, because of the transient nature of energy. The pattern, not the energy flowing through, has a certain degree of permanence. A structure must therefore be continually regenerating in order to be detected.

Structure is "local" because it is finite; it has a beginning and an end. "We cannot have a total structure of Universe" (606.01).

"Interacting" signifies the emphasis on relationships.

The phrase "complex of events" suggests an analogy to constellations, whose components—though spectacularly far apart—are interrelated for some cosmic span of time, creating a set of relationships, in other words, a pattern. The seven stars of the Big Dipper are light-years apart—the epitome of nonsimultaneous energy events. They are only perceived as a meaningful pattern from a special vantage point, Spaceship Earth. The remoteness of individual atoms in any structure or substance—not to mention the distance between atomic constituents—prompted Fuller to write "one of the deeply impressive things about structures is that they cohere at all." There is nothing "solid" about structure.

What do all structures have in common that allows their coherence? Triangles. At the root of all stable complexes is nature's only self-stabilizing pattern.

No Fuller study is complete without an "inventory": a list, not of each and every "special case" example, but rather, of the types of

Table II[a]

Polyhedron	Number of Edges	Volume	Volume per 6 Edges
Tetrahedron	6	0.11785	0.11785
Octahedron	12	0.47140	0.2357
Icosahedron	30	2.1817	0.4363

[a] Comparison of results in third column:

$$0.2357 = 2 \times 0.11785,$$
$$0.4363 = 3.70 \times 0.11785.$$

categories. The task, then, is to combine our new working definition of structure with the earlier one of systems. That means triangles involved in a subdivision of Universe. The virtually unlimited variety of irregular triangulated enclosures are not to be included in this inventory; rather, we seek a list of symmetrical stable enclosures.

And now we can sit back, for our task is already finished. Remember that only three systems can be made out of regular triangles: tetrahedron, octahedron, and icosahedron. These are the "three prime structural systems of Universe."

What can be learned through this kind of simplification? As in Euler's identification of vertices, edges, and faces, such categories organize the otherwise indigestible data to reveal new important features. An example is seen in Fuller's "structural quanta": the total material used for each of the three structural systems (easily measured in terms of number of edges) goes from six sticks to twelve to thirty. Chapter 10 will cover Fuller's ideas on the subject of volume in detail, but for now we can demonstrate an interesting fact while utilizing the traditional formulae of high-school geometry.

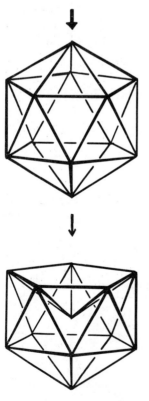

Fig. 5-4. "Dimpling."

Going from the smallest to the largest structure, the volume increases, not only absolutely, but relative to the number of edges. In other words, the ratio of volume to structural investment is a significant variable, increasing with additional structural quanta. The same holds true for ratios of volume to surface-area, as will be seen below. For clarity, we adopt unit edge lengths for all three polyhedra. Appendix B shows each step of the calculations, but the relevant results are displayed in Table II.

The implications of this information are suggested by Fuller's summarizing statement:

> The tetrahedron gives one unit of environment control per structural quantum. The octahedron gives two units of environment control per structural quantum. The icosahedron gives 3.7 units.... (612.10)

Fuller referred to six edges as a "structural quantum" because the total number of edges in each polyhedron is a multiple of six. "Environment control" simply refers to the ability to enclose and thereby regulate space.

Toward the goal of maximal enclosed space with minimal structural material (whether in terms of total strut length or surface area), designs based on the icosahedral end of the spectrum are advantageous. Hence Fuller's geodesic dome. For resistance to external loads, the tiny pointed tetrahedron is least vulnerable, for its concentrated structural elements resist buckling. The tetrahedron is all edges, enabling maximal structural resistance, and therefore highly applicable to truss design. (See Chapter 9.)

The icosahedron "dimples" easily. Fuller's term means just what it says. Push hard on one vertex and five triangles cave in, such that the tip of the inverted pyramid reaches just beyond the icosahedron's center of gravity. (See Fig. 5-4.) The tetrahedron is unique in being

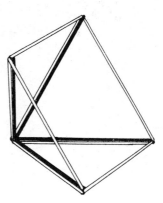

Fig. 5-5. Dimpling: one half of octahedron caves in to nest inside other half.

impervious to dimpling. Push hard on any vertex and either the whole system turns inside out (if the tetrahedron is made of rubber) or nothing happens; structural resistance prevails.

The octahedron, as expected, falls in the middle on both counts, that is, in terms of volume efficiency and load resistance. It will "dimple," but in so doing one half caves in to "nest" exactly inside the other half (Fig. 5-5).

Three of Fuller's inventions, the geodesic dome, the Dymaxion Map, and the Octet Truss, stem directly from the above principles. All three will be discussed in detail later, as other relevant geometric principles are revealed.

6

Angular Topology

Our study so far has primarily examined the conceptual foundation of synergetics. Except for occasional reference to volume and symmetry, the emphasis has been on numbers of elements rather than on shape. It's now time to look at the rest of the picture. Fuller's appreciation of the MIT definition of mathematics ("the science of structure and pattern in general") led him to ponder the appropriate tools and methods. "Science" is a systematic endeavor, requiring exact procedures for its description of structure and pattern.

A coordinate system describes the shape and location of a body in space by specifying the position of a sufficient number of that body's components. But a position can only be specified by its relationship to some other known location, or coordinate-system *origin*. In essence, mathematics functions by locating points relative to an agreed-upon frame of reference, such that the mathematician can say there is a point here and a point there and they are related by this kind of trajectory, and so on, until there is enough information to describe the entire system. Fuller points out that this information can be broken down into two aspects: shape and size.

What does "shape" consist of? "Shape is exclusively angular" (240.55): a simple but powerful observation. It's easy to envision identical shapes of completely different scale: for example, an equilateral triangle is a precisely defined concept, yet it contains no indication of size. It may be two miles or two centimeters in edge length, but its angles must be 60 degrees. Shape is influenced only by angle, and "an angle is an angle independent of the length of its sides" (516.02). The word "triangle" by itself (without further modification) does not describe a specific shape but rather a concept—three interrelated events without specific length *or* angle.

What does size consist of? Measurement, or dimension. In synergetics, these parameters are always expressed in terms of "frequency." The word is aptly applied, serving as a reminder of the role of time. Fuller dwells on the point: every real system ("special case") involves time and duration.[1] Real systems are events, and it takes time for an event to occur. He bases his objection to purely static concepts in mathematics on the fact that they are incompatible with twentieth-century scientific thought:

> Since the measure of light's relative swiftness, which is far from instantaneous, the classical concepts of instant Universe and the mathematicians' instant lines have become both inadequate and invalid for inclusion in synergetics. (201.02)

Since Einstein, Bucky reminds us, we can no longer think in terms of an "instant Universe," that is, a single-frame picture. Because even light has been found to have measurable speed, every aspect of physical Universe from the smallest tetrahedron to life itself involves the passage of time. Quite simply, "it takes time to get from here to there."

Frequency and Size

He insists upon nothing more adamantly than this distinction—between real ("experimentally demonstrable") phenomena and imaginary concepts. "Size" relates to real, time-dependent systems, whereas "shape," influenced only by angle and therefore independent of time, is a factor in both real and conceptual systems. ["Angles are . . . independent of size. Size is always special-case experience" (515.14).]

But how does "frequency" apply to size and length? Frequency connotes number: the number of times a repeating phenomenon occurs within a specified interval—ordinarily an interval of time, but Fuller extends the concept to include space. Length is measured in synergetics in terms of frequency to underline the fact that the "distance from here to there" involves time and can be specified in terms of number: number of footsteps across the room, or number of heartbeats during that interval, number of water molecules in a tube, number of inches, number of photons, number of *somethings*. The choice of increment depends on what is being measured, but frequency (and hence size) is inescapably a function of time and number.

Units of Measurement

Fuller explains frequency as subdivisions of the whole, suggesting another advantage of the term: it provides a built-in reminder that there is no absolute or single correct unit of measurement; rather, distance is measured relative to arbitrarily devised units. It is not a minor challenge to perceive distance this way; our conventional units —like inches or miles—are such an integral part of awareness that they seem a priori elements of size. The teacher in Fuller will not let us accept such useful conventions blindly, and so he employs tools such as "frequency" to keep us on our toes—aware of the nature of distance.

Time and Repetition: Frequency versus Continuum

Just as length cannot exist without time, there also could be no awareness without time. Time, inseparable from all other phenomena, cannot be isolated. "Time is experience" (529.01).

The concept of time is inextricably tied to awareness; appropriately it is measured in terms of the frequency of detectable repeating events. Periods of daylight reliably alternating with darkness gave us a unit we call a "day." Heartbeats might have defined the "second," planting the awareness of that tiny increment in long-ago human beings. The predictable repetition of days growing longer and shorter with their accompanying weather changes defined a "year." To conceive of time requires repetition.

However, the limits of perception prevent recognition of the periodicity in very high-frequency patterns such as light waves or repeating molecules in a toothpick. If repetition is too frequent, we perceive a continuum rather than segmented events. Fuller's use of "frequency" to specify size draws attention to the nonexistence of continuums. Here, as always, his goal is to develop a mathematical language which accurately represents reality.

Shape and size are thus replaced by angle and frequency.

Fuller's *principle of design covariables* summarizes by stating that two factors are responsible for all variation. "Angle and frequency modulation exclusively define all experiences, which events altogether constitute Universe" (208.00). In short, "structure and pattern in general" are described completely by only two parameters: angle and frequency—another way of saying that the differences between systems are entirely accounted for by changes in angle and length. Again the goal of such simplification is the demystification of mathematics.

Remember that Fuller's overall goal was to isolate "nature's coordinate system"—by which he meant the simplest and most efficient reference system to describe the events of nature. We gradually narrow in on his solution.

Topology and Vectors

Fuller has declared his scope: "Synergetics consists of topology combined with vectorial geometry" (201.01). Topology in essence analyzes numbers of elements. (Euler's law is topological, involving neither symmetry nor size.) And now we must again think about vectors, for they are the key to this combination.

Vectors provide an ideal tool for representing velocity, force, and other energetic phenomena. As you may recall, the concept is actually quite simple—despite its lack of popularity among high-school students. A vector is a line with both specific length and angular orientation. It's the ultimate simplification of actions or forces, presenting only the two most relevant bits of information: magnitude and direction. Mathematics defines this tool and the accompanying rules for its manipulation, just as it defines the set of real numbers and the rules for addition and multiplication. The mechanism as a whole enables us to predict the results of complex interactions of forces and bodies in motion.

Surveying classical geometry, Fuller decided that "there was nothing to identify time, and nature has time, so I'd like to get that in there." Vectors seemed to provide the solution, "I liked vectors. A vector represents a real event of nature .... I wondered if I couldn't draw up a geometry of vectors; that would mean having the elements of experience."[2]

Back to synergetics: What is meant by a combination of topology and vectorial geometry? And how does it fit into the search for "nature's coordinate system"? By viewing polyhedra as vector diagrams, Fuller integrates the two subjects (vectors and topology) in a deliberate attempt to develop one comprehensive format to accommodate both the inherent shape of space and the behavior of physical phenomena. Polyhedra with vectors as edges necessarily incorporate both shape and size.

Vector Polyhedra

The spectrum of possible forms of polyhedra is certainly informative about the shape of space; polyhedra are systems of symmetry made visible. Any configuration allowed by space can be demonstrated by vertices and edges, and, as noted earlier, experimentation quickly reveals that the variety of possible forms is limited by spatial constraints. Furthermore, the shape of space is fundamental to the events of nature. Fuller believed that mathematics, the science of structure and pattern, should be based on these principles.

So Fuller coined the rubric "angular topology" to express what he saw as the principal characteristic of synergetics: integration of the static concepts of geometry with energetic reality. These may not have been the words he used back then, but the desire for such a system dates back to Bucky's early school days. Or at least that's how the story goes. Such myths evolved over time to convey the spirit of the child's inquisitive confusion through concise anecdotes. Fuller's lectures and writings incorporate a full repertoire of autobiographical moments in which the young Bucky has startlingly complete and rich realizations. The process is beautifully described by Hugh Kenner, who relieves us of the burden of asking, "Did that really happen, just like that, one morning"?

Not that he deceives. He mythologizes, a normal work of the mind... to embrace multitudinous perceptions, making thousands of separate statements about different things [into] summarizing statements...

What a myth squeezes out is linear time, reducing all the fumblings and sortings of years to an illuminative instant. We can see why Bucky needs myth. The vision that possesses him eludes linearity... The myth is anecdotal.[3]

We return to the geometry lesson: the grade-school teacher has put a drawing on the blackboard and said, "This is a cube." Young Bucky wonders aloud, "How big is it, how much does it weigh, what is its temperature, how long does it last"? The teacher says, "Don't be fresh," and "You're not getting into the spirit of mathematics." Again, the implication is that mathematics does not deal with real things, but only with absurd constructs and arbitrary rules.

It's not hard to accept the message behind the story—that something about the teacher's lesson was profoundly disturbing to the child. It seemed to Bucky that mathematics was a serious enterprise and it should limit itself to "experimentally demonstrable" phenomena. That meant no fooling around in the fringe area of sizeless points, infinite planes, and weightless cubes. We reconsider these early musings in the new context of "angular topology" to see where they led, and recall that Fuller was later to attribute mathematics' lack of popularity to the perfectly natural discomfort people felt with those elusive concepts. Explanations ought to be in terms of tangible experience.

But then what are we to make of such claims as "Synergetics permits conceptual modeling of the fourth and fifth arithmetic powers; that is, fourth- and fifth-dimensional aggregations of points or spheres in an entirely rational coordinate system that is congruent with all the experimentally harvested data of astrophysics and molecular physics..." (202.01)? Under the heading "Angular Topology," this statement is found too early in Fuller's *Synergetics* to be easily understood. One might wonder if a page was left out of that particular copy; but the root of the confusion is not that easily located. With some additional background material, we can begin to understand Fuller's assertion. The word "dimension" has been lurking behind the scenes in this entire discussion, and must now be brought out into the open.

Dimension

(1) A measure of spatial extent. (2) Magnitude, size, scope. (3) The number of factors in a mathematical term. (4) A physical property, often mass, length, time, regarded as a fundamental measure. (5) Any of the least number of independent coordinates required to specify a point in space uniquely.

The above is a sampling of what you will find in English-language dictionaries under "dimension." As you can see, there are a few distinct meanings—essentially falling into three categories:

(1) Physical extent or measurement, as in "what are the dimensions of this room"?
(2) Orders of complexity, in the most general sense, as in "the many dimensions" of an issue or problem. This meaning is as common as it is widely applicable.
(3) The specifically mathematical application: the number of independent terms required to specify a point in space. Our conventional system utilizes three independent, mutually perpendicular axes in space to accomplish this task. This is often the first meaning to occur to people, especially when already thinking about geometry. "Space is three-dimensional." However, this assignment—treating the third category as an exclusive definition—seemed unacceptably limited to Fuller. Exposure to the ordered polyhedra of mathematics and also to organic structures and crystals found in nature makes an orientation toward perpendicularity seem quite arbitrary. Although right angles are sprinkled throughout geometric shapes, they are by no means dominant. And, more often than one might expect, ninety-degree coordinates provide an awkward framework with which to describe both naturally occurring and conceptual formations.

Fuller viewed the Cartesian coordinate system with its three perpendicular axes, conventionally labeled X, Y, and Z, as a remnant of "flat-earth thinking." Early man, finding himself on a huge flat expanse, assumed that up-and-down and back-and-forth were the fundamental directions of his universe. Ninety degrees was the obvious natural angle with which to segment and measure space. Humankind has had more and more evidence of nature's radial and spherical bias throughout history—from the discovery of the shape of planets to the behavior of radiation and cellular growth. But neither Copernicus's spherical earth nor the vast array of biological and physical phenomena, all suggesting that angles other than ninety degrees would provide more "natural" or convenient standards, succeeded in reorienting the perpendicular bias of mathematics.

The "three dimensions" of mathematics—length, width, and height—became part of an unshakable convention. That space cannot accommodate a fourth perpendicular direction is just one of its many intrinsic constraints, and yet this limitation is too often seen as the only characteristic of space. While mathematicians postulate hypothetical "hypercubes" in their attempt to describe a spatial fourth dimension, and physicists refer only to "time" as the fourth dimension, Bucky preferred to call attention to the "four-dimensional" *tetrahedron*. Time is certainly a dimension, but the physicists' progression "x, y, z, and t" seemed not to emphasize sufficiently that time—permeating all space and all experience—is qualitatively unlike the other "three dimensions."

As we develop an awareness that space has shape, right angles gradually seem less "natural." The *XYZ* coordinate system often serves to obscure rather than illuminate spatial characteristics. It is a valuable tool, which we can recognize as one alternative superimposed by human minds, not as a framework organic to the shape of space itself. The word "dimension" is used without contradiction to describe the maximally symmetrical arrangement of three lines in space; likewise it can be applied to time, but it's not the end of the story.

One of the above dictionary definitions refers to the number of coordinates required to specify the location of a point in space. Assuming the existence of a previously specified origin, the number of coordinates happens to be three. Does this result reinforce the exclusive use of the *XYZ* axes? No, for the three coordinates required do not have to be Cartesian; another option is *spherical coordinates*, in which the location of any point is fixed by specifying two angles and a radial distance. Cartesian coordinates, on the other hand, describe a location as the intersection of three lines originating at given distances along three perpendicular axes. (See Figure 6-1 for

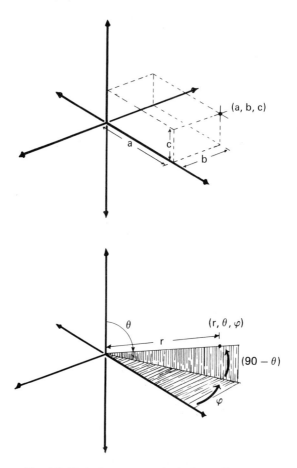

Fig. 6-1. Cartesian versus spherical coordinates.

a comparison of the two methods.) The spherical approach is more suited to Fuller's radial ("converging and diverging") Universe; its emphasis on angular coordinates encourages thinking in terms of "angle and frequency modulation."

Dimension is a widely encompassing term, and can legitimately refer to numbers of factors in a variety of geometric phenomena. Considerable time can be devoted to unraveling Fuller's different uses of "dimension" in *Synergetics*, and we shall continue to cite examples throughout our investigation.

Size

Fuller's book takes a firm stand in the opening sections: "Synergetics originates in the assumption that dimension must be physical" (200.02), meaning *size*. The declaration is soon reinforced: "There is

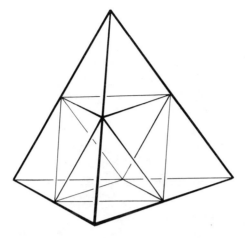

Fig. 6-2

no dimension without time" (527.01). Firmly imbedded in reality now: it takes time to embody a concept. Everything ties together, so far.

It would be uncharacteristically clear-cut if that were Fuller's only use of dimension. Synergetics may start with dimension as size, but other applications of the multifaceted term are sprinkled throughout the book. (Identification of space as three-dimensional is not one of them.) Keep in mind that this mathematical convention has a firm hold; it's difficult to think otherwise about space—and consequently not easy to view Fuller's material objectively.

Planes of Symmetry

Both the tetrahedron and the octahedron—two of the simplest structures—incorporate four nonparallel planes. The faces of the tetrahedron present four distinct directions, just as the faces of the cube provide three. The octahedron has four pairs of opposing parallel triangles, and it can be demonstrated that they are parallel to the tetrahedral faces (Fig. 6-2). Fuller refers to this geometric trait as "dimension," and through repeated observation places considerable emphasis on the inherent fourness of the "minimum system in Universe." "The octahedron's *planar* system is four-dimensionally referenced, being parallel to the four symmetrically interacting planes of the tetrahedron..." (527.31). The icosahedron, on the other hand, exhibiting various fivefold symmetries, embodies "five-dimensionality" (527.50) in Fuller's unorthodox appropriation of terms.

Other Applications of Dimension

Another twist: Fuller also refers to the three structural parameters—vertices, edges, and faces—as different dimensions of structure. In a later section covering the concept of dimension, Fuller reintroduces "constant relative abundance" (as explained in Chapter 4) under the heading "527.10 Three Unique Dimensional Abundances"—namely, vertices, edges, and faces. This and other ambiguous—if not contradictory—usages of certain terms can often obscure the mathematical statement being made. In this case, Fuller's point about the consistent arithmetic relationships between vertices, edges, and faces in closed systems is lost amid confusion about the meaning of "dimensional abundances."

Another unorthodox usage involves pairs of opposites. At one point in *Synergetics*, for example, a magnet, with its positive and negative poles, is called a two-dimensional system. Along the same lines, "Polar points are two dimensional: plus and minus, opposites" (527.21).

Finally, "dimensional aggregations" in the opening quotation of this section refers to numbers of layers in certain clusters of closely packed spheres. We shall explore these patterns in Chapter 8. Fuller's different uses of "dimension" may be confusing, but they are not, strictly speaking, incorrect—at least not in terms of the dictionary. Mathematical convention is another issue.

Fuller does mention the historical precedent for conceiving of space as exclusively "three-dimensional," thereby explaining his license to reevaluate our concept of dimension; however, the reference is too late in the book to clear up early confusion:

> ...The Greeks came to employ 90-degreeness and unique perpendicularity to the system as a basic...dimensional requirement for the...unchallenged three-dimensional geometrical data coordination. (825.31)

So, while he does justify his usage with this reference to the word's flexibility, the clarification is obscured by the book's sequence. The reader seeking a quick reason to dismiss *Synergetics* might focus on Fuller's extravagant citation of other dimensions early in the book. His apparent familiarity with "the fourth dimension" provides just cause for suspicion; however, a simple change of article—from "*the*" to "*a* fourth dimension"—gives the term a very different effect.

Like most subjects in synergetics, "dimension" cannot be neatly presented in one complete package; boundaries are never that clearly defined. In addition to the fact that different subjects overlap, there can always be new twists. The trick is to leave ourselves open to exploration, free to evaluate each new application without bias.

Angular Topology

Once in a long while, a "generalized principle" takes recognizable shape and emerges out of the vast sea of man's cumulative findings. For Fuller, these principles—characterized as true in every case—are the real wealth of society. Applications may not always be immediately clear, but if an inventory of "generalized principles" is made accessible, he reasoned, humanity will put them to use sooner or later.

The "principle of angular topology" was recognized by the mathematician and philosopher René Descartes (1596–1650), but the title of course is Fuller's. Perhaps by giving Descartes's remarkable discovery a new title, Bucky hoped to excite the kind of attention he felt it deserved.

In every polyhedral system, the sum of the angles around all the vertices is exactly 720 degrees less than the number of vertices times 360 degrees, or $(360° \times V) - 720°$. True for the tetrahedron, true for the crocodile. In Fuller's words, every system has exactly 720 degrees of "takeout."

If this principle seems complicated, it is only because the words are hard to follow, but the following image should make it easier. Picture a paper cone—the shape of an ice-cream cone without the ice cream. Notice that a cone, having an opening at the base, is not a closed system. Now, split the paper cone open by slicing a straight line from its pointed tip to the circular hole, and then spread the piece of paper out flat on the floor like a rug (Fig. 6-3). There is now an angular gap left by the paper, where the floor shows through. That gap is the "takeout angle," the angular difference between a flat map and a cone.

In the same manner, you can slice open some number of edges of a polyhedron until *its* surface can be spread out like a rug. The resulting map, similar to a dressmaker's pattern, is called in geometry a polyhedron's *net*. A net contains all faces of a polyhedron, some of them separated by angular gaps; it is a flat pattern which can be folded along the edges, and taped together to generate its polyhedron. Figure 6-4 shows nets of an icosahedron and an oc-

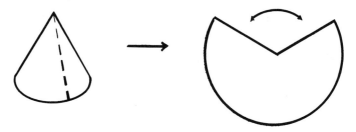

Fig. 6-3

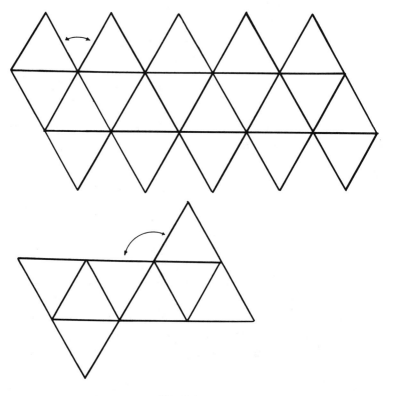

Fig. 6-4

tahedron. The principle of angular topology states that the sum of the angular gaps, no matter how simple or complex the system, will be exactly 720 degrees.

Or go through these procedures in reverse: start with flat paper and cut out one pie-shaped segment to make a cone. Then keep going, cutting out more slices, just until the point at which the paper can be closed off into a system. This point is reached when you have taken out exactly 720 degrees. It's a prerequisite to closure; there is no leeway. To check, you can measure all the scraps that have been taken out; the results are always the same: however irregular your cuts or strange your resulting closed system, the total takeout must be 720 degrees. This consistent total presents a generalized principle for closed systems.

The surface angles of any tetrahedron (regular or not) also happen to add up to 720 degrees. (Four triangles: $4 \times 180° = 720°$.) Fuller certainly isn't going to let that one slip by! The "difference between the visibly definite system and the invisibly finite Universe [i.e., plane] is always exactly one finite invisible tetrahedron..." (224.10).

Consider once again the variety of systems. This principle—a first cousin of Euler's law—describes an extraordinary consistency. The "720-degree excess" is an appropriate parallel to Euler's "constant 2" in that there are 720 degrees in two complete revolutions. Both are counterintuitive: Euler's law reveals that the number of edges is always exactly two less than the vertices plus faces, no matter how complex the system, just as the angular "takeout" is 720 degrees whether the surface angles themselves add up to a total of 720 degrees, as in the tetrahedron, or to 57,600 degrees, as in the "four-frequency icosahedron." (Don't worry, that structure will be explained below.) Table III shows the results for a few different polyhedra, verifying the constant "excess" of 720 degrees.

Table III reveals another notable consistency: the sum of the surface angles in every polyhedron is a multiple of the tetrahedron's 720 degrees (column 5). This calls to mind our earlier observation that the number of edges in many ordered polyhedra is a multiple of the tetrahedron's six. (Refer to Chapter 4.)

Table III

Polyhedron	1 Number of vertices, V	2 V × 360°	3 Sum of surface angles	4 (Col. 2) −(Col. 3)	5 (Col. 3) /720
Tetrahedron	4	1,440°	720°	720°	1
Octahedron	6	2,160°	1,440°	720°	2
Icosahedron	12	4,320°	3,600°	720°	5
Cube	8	2,880°	2,160°	720°	3
Pent. Dodec.	20	7,200°	6,480°	720°	9
VE	12	4,320°	3,600°	720°	5
4f Icosa	162	58,320°	57,600°	720°	80

Angular Takeout: An Example

A complicated system such as the four-frequency icosahedron provides an especially good illustration of this remarkable consistency. The structure is an irregular polyhedron with 320 triangular faces, and is based on the symmetry of the icosahedron. Each icosahedral triangle is replaced by sixteen new smaller triangles, producing the total of 320 faces of this more or less spherical structure. Chapter 15 will describe the origin of high-frequency icosahedral enclosures in detail, but for now, we can understand that the faces are irregular triangles. As shown in Figure 6-5, most vertices join six triangles, and we recall from Chapter 4 that if six *sixty-degree* angles meet, they create a flat surface. Therefore, if six faces are to surround a convex vertex of a polyhedron, their angular total must be less than 360 degrees—which produces the "angular takeout". Those interested in exploring how to calculate individual edge lengths and surface angles can refer to Appendix A, "Chord Factors," and to Appendix C for a list of additional sources; other readers should simply be aware that the values will be highly irregular numbers. Having noted that, to assure convexity, the surface angles must add

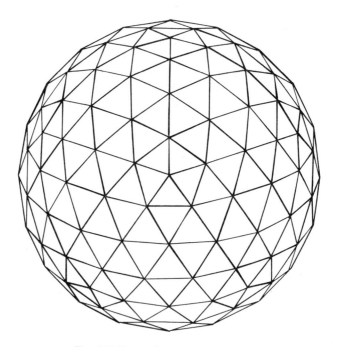

Fig. 6-5. Four-frequency icosahedron.

up to less than 360 degrees at each vertex, we can reflect on how extremely small the gaps at each vertex in this structure will be.

If all 162 vertices are equally distant from the structure's center, the average total of the surface angles at each will be 355 degrees, 3 minutes, and 20 seconds—or 355.5556 degrees. Thus, the "takeout" angle—instead of being concentrated at a few points, with 60 degrees removed from each of twelve vertices— is distributed among many component triangles. Imagine cutting open this system and spreading out its multifaceted net. Each angular gap would be very narrow, averaging 4 degrees, 26 minutes, and 40 seconds (4.4444 degrees.) If we were to draw this net, the pencil thickness itself would be a nuisance. Supposing each edge is approximately one inch long, the outermost or widest part of each gap will be less than one-twelfth of an inch.

In addition to being miniscule, these numbers are typically quite irregular, not at all simple fractions of degrees. Nevertheless, these gaps, calculated (say) to six decimal places, add up to exactly 720.000000 degrees, no matter how many vertices in the system, or how irregular the distribution.

One interesting implication of the principle of angular topology is further demonstration of the impossibility of the traditional sphere as defined by mathematics—an unreachable planar 360 degrees around every one of an infinite number of vertices. "The calculus and spherical trigonometry alike assume that the sum of the angles around any point on any sphere's surface is always 360 degrees" (224.11). Fuller goes on to point out that in order to achieve a closed system, there must be 720 degrees taken out, distributed throughout the vertices, thereby invalidating this assumption. "The demonstra-tion thus far discloses that the sum of the angles around all the vertexes [*sic*] of a sphere will always be 720 degrees—or one tetrahedron—less that the sum of the vertexes times 360 degrees—ergo one basic assumption of the calculus and spherical trigonometry is invalid" (224.11). In other words, since the 720-degree takeout is a prerequisite to closure, even a sphere has to have infinitesimally less than 360 degrees around any given point on its surface. (We are forced to conclude that "infinitesimal" times "infinite" here equals 720 degrees.)

Angle Types

Finally, knowing the different types of angles in geometry will be helpful. The nomenclature is straightforward. *Surface angle* is by

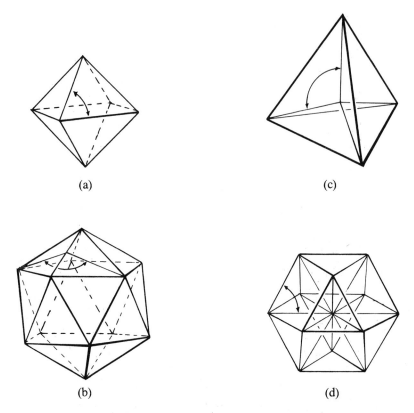

Fig. 6-6. (a) Surface angle; (b) dihedral angle; (c) central angle; (d) axial angle.

now familiar, referring to a corner angle of a polyhedral face (Fig. 6-6a).

Dihedral angles are the angles between adjacent faces, on the *inside* of a system. (Fig. 6-6b).

A *central angle* corresponds to a polyhedral edge and is measured from the exact volumetric center of a system to each end of the edge (Fig. 6-6c). Central angles provide a way to find exact edge lengths of a given system for any desired radius.[4] Central angles provide an effective way to record *relative* lengths, for remember, an angle is independent of the lengths of its sides. This means that we can list the complete set of edge lengths for a complex polyhedral or geodesic system in terms of the central angles corresponding to each edge, and then directly calculate exact lengths for any desired radius, with the help of a simple trigonometric equation. This process might sound complicated at first, but is more expedient than the alterna-tive, which is to specify *one* set of edge lengths, applying to only one "special-case" system. Data for any different size would then have to be completely recalculated, step by step, from scratch. Central angles give us data for the general case, applicable to any particular realization of the same shape. In architecture, this allows us to build a geodesic dome of any size from one set of central-angle calculations. A pocket calculator is all that is required to simply multiply the desired radius by twice the sine of half the central angle.[4]

An *axial angle* is the angle between the edge of a polyhedron and an adjacent radius (Fig. 6-6d).

That finishes this chapter, but the subject of angle is never far removed from any discussion in synergetics.

7

Vector Equilibrium

If you begin to suspect that the concepts hiding behind Fuller's intimidating terminology are often easier to understand than their titles, you will soon find that "vector equilibrium" is no exception. "The vector equilibrium is an omnidirectional equilibrium of forces in which the magnitude of its explosive potentials is exactly matched by the strength of its external cohering bonds" (430.03). If Fuller's description doesn't make it crystal clear, read on! The VE, as it is usually called, is truly the cornerstone of synergetics.

Vectors are certainly familiar to us by now; but what is meant by equilibrium? The word is by no means esoteric; like "systems," it enjoys considerable popularity these days. That is no wonder, for the concept is at the root of all phenomena, both physical and metaphysical. Defined by *The American Heritage Dictionary* as "any condition in which all acting influences are cancelled by others, resulting in a stable, balanced, or unchanging system," equilibrium is not inactivity, but rather a dynamic balance.[1] This balance is not necessarily physical, but may be mental or emotional as well. In fact, so much of experience is characterized by fluctuation in and out of fragile balances, that it is easy to understand the word's frequent use —covering everything from structural to emotional to financial equilibrium.

A simple mechanical model of equilibrium involves a ball and a bowl. Allowed to roll around inside the open smooth surface, a ball will finally come to rest at the bottom of the bowl, requiring renewed force to set it back in motion. This state is called *stable equilibrium*. On the outside of an inverted bowl (or dome) the ball might rest briefly at the center, but the slightest disturbance will make it roll off —thus demonstrating *metastable equilibrium*. The third possibility is that the ball sits on a flat table, in a state of *neutral equilibrium* (Fig. 7-1).

Nature exhibits a fundamental drive toward equilibrium. Scattered pockets of varying temperatures will equalize at the mean tempera-

Fig. 7-1. Stable, metastable, and neutral equilibrium.

ture; opposing forces of different magnitude naturally seek a state of rest; these differences cannot remain imbalanced. Greater forces overpower smaller forces, causing motion until they balance out. Demonstrations of this universal tendency are provided by countless everyday experiences. For example, if a massive object sits on too weak a shelf, the force exerted by gravity toward the earth's center exceeds the strength of the shelf's restraining force and the object comes crashing through. Motion continues until a new equilibrium is achieved by the object landing on a sturdy floor capable of matching the gravitational force with an equal and opposite restraining force. Apparent motion then ceases, as a stable equilibrium is maintained.

Invisible motion continues, however; atoms never stand still. The systems approach encourages us to note that we can zoom in to observe the same event on another level of resolution.

The front door is opened and quickly closed, allowing a rush of cold winter air into the living room. Freezing temperatures dominate the corner of the room near the door, while the other side by the radiator is cozy and warm. However, the imbalance quickly disappears; the temperature soon becomes more or less consistent throughout the room.

Nature's tendency to seek equilibrium is a spontaneous reaction; it is the path of least resistance.

We have already seen that vectors model certain events and reactions of nature. In this discussion, we focus on one specific use of vectors: to represent forces. The application is straightforward. Forces push or pull on something. The strength or *magnitude* of a force is represented by the length of the vector, and its *direction* is of course specified by the orientation: frequency and angle, as Bucky says.

It follows, then, that a balance of forces is geometrically modelable. We can create a spatial diagram of the concept of equilibrium, and in so doing learn more about space's inherent symmetry.

Bucky's love affair with vectors dates back to his World War I Navy experience. Introduced to vector diagrams of colliding ships in the officer's training program, he discovered that the tiny arrows contained all the necessary information about the ships' masses and speeds and compass headings to predict the results of collisions or the effect of tail winds and other influential forces. Bucky was fascinated by the economical elegance of the system. These vectors actually modeled the energetic events of reality—a pleasant contrast to the mathematics teacher's "lines stretching to infinity." Bucky was hooked. "A vector is an experience," he reminisces in a 1975 videotaped lecture, "so I thought, if I could only have a geometry of vectors, that would be great."[2] This concept was introduced in the previous chapter, but now we must discern the specific shape of the configuration generated by vector diagrams and models.

"Nature's Own Geometry"

We periodically remind ourselves of the purpose behind this geometric journey. Trying to faithfully trace Bucky's footsteps, we seek to isolate the "coordinate system of nature": how Universe is organized. One of the essential parts of the mystery is how to account for structural similarities between totally unrelated phenomena, vastly different in both scale and material. The implication is that, rather than *being* coordinated, things coordinate themselves. This self-organization occurs according to a set of physical forces or constraints, absolutely independent of scale or specific interactive forces. In short, space *shapes* all that inhabits it.

But how? Through what vehicles does nature adhere to this underlying order? Let's look at Fuller's fundamental operating assumption:

> It is a hypothesis of synergetics that forces in both macrocosmic and microcosmic structures interact in the same way, moving toward the most economic equilibrium packings. By embracing all the energetic phenomena of total physical experience, synergetics provides for a single coherent system of geometric principles. ... (209.00)

Synergetics seeks to establish the natural laws through which the self-organization of systems in the most diverse fields of science occurs. Science, as we noted earlier, acknowledges a fundamental drive toward equilibrium, but what else can be observed about this tendency?

Gas molecules buzzing around in a closed-container are suddenly allowed, by the removal of a barrier, into an adjacent empty compartment of the same size (Fig. 7-2). The molecules rapidly disperse, taking advantage of their new freedom by using the additional room

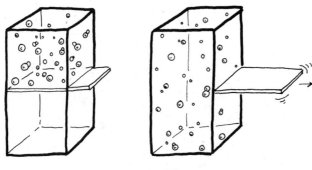

Fig. 7-2

to spread out and slow down. The reverse action—of all the gas molecules suddenly gathering in one half of a container—has never been observed, just as in the living room disparate temperatures equalize, but that room will never spontaneously become warm on one side while the other side suddenly cools off. The closed-container experiment is the classic model for illustrating nature's entropic tendencies. The Austrian physicist Ludwig Boltzmann (1844–1906), noted for his work in entropy theory, called the resulting dispersal disorderly behavior. A geometer, however, might observe the individual gas molecules vying for the most room—accomplished of course by a maximally symmetrical distribution—and not perceive such a progression as disorder. Both perceptions call it equilibrium.

Spatial Considerations

A symmetrical distribution of "energy events" involves a large number of equivalent separation distances. For a more tangible image than provided by gas molecules, picture a large room full of people asked to spread themselves out for stretching exercises. If the room is sufficiently crowded, a more or less triangular pattern in the distribution of people can be observed, as a result of individuals' trying to maximize the area of their territory. Each person feels he or she has more space when the distances between people are maximized, which is the case when all distances are as close to equivalent as possible. If it's hard to see why that equivalence implies a triangular pattern, read on.

Think about baking cookies on a tray. Intuition rather than geometrical training tells you that the cookies in successive rows should be offset to maximize the number which can fit on each tray without spreading into each other. Observe in Figure 7-3 that a

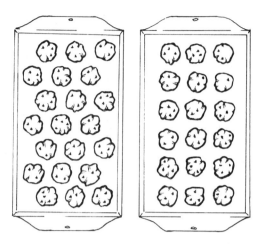

Fig. 7-3. Triangular pattern enables one more row of cookies than square pattern with the same minimum separation distance between cookies.

square distribution with the *same minimum separation* between cookies wastes considerable tray area, resulting in fewer cookies than a triangular pattern.

In the same way, people in a room naturally (and quite unconscious of the advantages of triangular distribution) milling around until each carves out a desirable comfort zone can end up by increasing the overall symmetry. This organization does not require a director at the head of the room. Nature behaves in the same manner, seeking the most comfortable resting position. Forces continue to push or pull until counterbalanced, and in the absence of other influences, symmetrical considerations dominate. (There is, in Fuller's words, an "a priori absolute mystery" of *why* nature behaves this way, which is beyond explanation. The question is thus *how* Universe operates.)

The advantageous balances suggested by the term equilibrium can be expressed in terms of symmetry. The properties of space are necessarily behind all events and reactions in nature; hence Fuller's assertion that forces in both "macrocosmic and microcosmic structures interact in the same way." Space is the same on every scale, embracing and molding the "most economic equilibrious packings."

With the conceptual foundation in place, we can now describe the model proposed by Fuller to represent equilibrium. In his words, we seek the simplest "omni-accommodative system" able to model the behavior of complex systems. Basically, we want to draw, or better yet, *build* that much-discussed balance of forces. To accomplish the desired result, a model must incorporate two aspects of Fuller's geometry: first it must consist of vectors, and secondly it must cover all directions. In short, we want to illustrate an equilibrium of vectors in space.

Planar Equilibrium

Spatial configurations tend to be difficult to visualize, whereas flat patterns are not, so we start with the page. Draw a vector of some arbitrary length—which we designate "unit length"—in any direction. To counteract that force, we position a second vector directly opposite the first, head to head (Fig. 7-4a). They have the same magnitude and opposite direction and are therefore balanced.

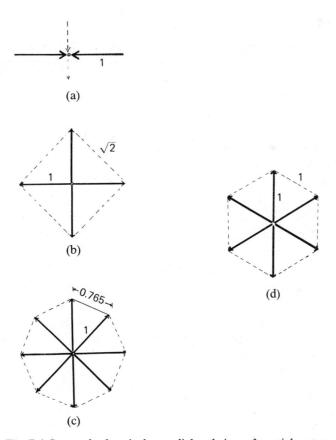

(a)

(b)

(d)

(c)

Fig. 7-4. In search of equivalent radial and circumferential vectors.

It is an unstable balance however, easily knocked out of equilibrium by a force from any other direction. Suppose that the original two forces push on a body with equal strength from east and west. A force from the north or south, even if smaller than the east–west pair, can easily destroy that unstable equilibrium. So how can we most efficiently insure the stability of the body in question?

Suppose instead that the forces are directed outwardly from the body (Fuller's "explosive potential," as quoted in the first paragraph of this chapter). To begin with, imagine four equal vectors, heading north, south, east and west—that is, in the positive and negative directions of the X and Y axes. To counteract the four explosive forces, we need equivalent restraining forces ("implosive" or "embracing"). Fuller's "embracing" vectors are not technically part of the conventional language of vectors. Having neither head nor tail, their effect is simply restraining, like a net, and their magnitude is still assumed to be represented by their length. As strength is graphically depicted by vector-length, we soon find that there is no easy way to draw four embracing vectors of unit length. In Figure 7-4b, the ends of our four explosive vectors are interconnected, but these new lines are approximately 1.414 times as long as the outward forces. The longer vector lines represent more powerful forces and thus overpower the explosive potential, meaning that the whole display must collapse inwardly.

We might then choose to add more outward forces, maintaining symmetry by an additional unit vector in between each of the original four (Fig. 7-4c). Now we have eight equal forces emanating from one point, and the resulting eight embracing lines are only 0.765 times as long as the unit length—too *small* to restrain the explosive forces. This imbalance leads to outward dispersal of the hypothetical system.

The sought-after balance will be achieved by "omnisymmetry", that is, maximum symmetry. The desired array must consist exclusively of unit-length vectors—both explosive and embracing.

One way to solve the problem is to picture a square grid of "energy events," interconnected by vectors. Squares provide an easy starting point because they make up the basic framework of current mathematics: the XY coordinate system. As before, the length of vectors which connect "events" represents the strength of their interattraction. Because the distance between diagonal corners is 1.414 times the distance between adjacent loci, the attractive forces represented are that much greater. Imbalance (or lack of equilibrium) in a diagram of forces represents motion. As a result of the attractive forces between neighboring energy events, they push and pull on each other until all the disparate separation distances became equal. Again, like the cookie tray, a pattern of equilateral triangles is established. Every single energy event is a uniform distance apart from each neighbor, and there are 60-degree angles between all vector connections. The forces are finally balanced, and the resulting array informs us about the fundamental symmetry inherent in a flat surface.

Going back to the radial vector diagram, we now arrange six vectors emanating from a point. The embracing vectors will be the same length as the "exploding" group. An inescapable consequence of this fact (obvious in retrospect) is that angles between all vectors are also identical—not just those between radial vectors, but also those between circumferential and radial vectors (Fig. 7-4d). No other arrangement has this property, because 60-degree angles are integral to equilateral triangles.

A diagram of radial vectors can be thought of as an apple pie divided into some number of pieces. These pieces are always triangular, with two radial edges and one circumferential edge. Because of the fact that the angular sum in every flat triangle is 180 degrees, the only way for a triangle to have all angles the same is with three sixty-degree angles. Therefore, the only angular measurement that will allow us to divide the vector pie "omnisymmetrically" is 60 degrees, requiring that unity (360 degrees) be divided six ways. The procedure is straightforward so far. In the plane, equilibrium is demonstrated by a hexagon (Fig. 7-4d).

Now we make the leap into space, with its accompanying leap in complexity. It may be difficult to visualize a spatial array, especially noncubical configurations, but taking it step by step, we shall be able to develop and understand Fuller's model.

In terms of vectorial dynamics, the outward radial thrust of the vector equilibrium is exactly balanced by the circumferentially restraining chordal forces: hence the figure is an equilibrium of vectors. All the edges of the figure are of equal length, and this length is always the same as the distance of any of its vertexes from the center of the figure. (430.03)

A "geometry of vectors," Fuller reasoned, must be "omnidirectionally operative"—hence, radially oriented and omnisymmetrical. Following the planar example, we want some number of vectors emanating from an origin, situated so that the distances between vector end points (vertices) are not only all equal to each other, but also exactly equal to the length of the radial vectors.

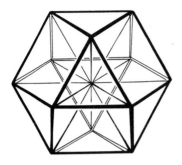

Fig. 7-5. Vector equilibrium.

Cuboctahedron as Vector Equilibrium

We can understand the symmetry of the plane by observing that although any polygon can be made to have equal edge lengths, only the regular hexagon can have edges equal in length to the distance between the polygon's center and its vertices. In the same way, although there are many regular and semiregular polyhedra with equal edge lengths, there is only one spatial configuration in which the length of each polyhedral edge is equal to that of the radial distance from its center of gravity to any vertex: the cuboctahedron (Fig. 7-5).[3] This shape therefore is the only one that allows the requisite arrangement of vectors to demonstrate equilibrium.

We first saw the cuboctahedron as the degenerate truncation of both the cube and the octahedron, but at that point in our investigation we were only looking at surface topology. Now diving into the interior shape, we discover this unique property of equivalence. Table IV compares the radial lengths of various familiar polyhedra given unit edge lengths. Only in the cuboctahedron—hereafter referred to by Fuller's term, vector equilibrium or VE—can the radius be of unit length.

Again, in order for all vectors to be exactly the same length, the angles between them—both radial and circumferential—are necessarily equal. In Figure 7-5, the VE is shown with both radial and

Fig. 7-6. (a) Cuboctahedron; (b) twist cuboctahedron.

Table IV

Unit-edge polyhedron	Radius	Central Angle	Axial Angle
Tetrahedron	0.6124	109.47°	35.26°
Octahedron	0.7071	90.00°	45.00°
Icosahedron	0.9511	63.43°	58.28°
Cube	0.8660	70.53°	54.76°
Pent. dodecahedron	1.4012	41.81°	69.04°
VE	1.0000	60.00°	60.00°

edge vectors. Radial vectors connect the twelve vertices to the system's center, thereby forming twenty-four radiating equilateral triangles, corresponding to each polyhedral edge and pointing inwardly. We should not be surprised to find an array of equilateral triangles in the VE, for this is the only polygon with equal distances and angles between all points. And, as vectors incorporate both magnitude and direction, an equilibrium of vectors must—in Fuller's terminology—balance both angle and frequency. Sixty-degree angles are inevitable.

What if we had anticipated the necessity of 60-degree angles? We could then have *started* this part of the investigation by specifying the angle between radial vectors and looking for the resulting implications of that choice. The discovery then would be that the necessary 60-degree gaps in a spatial array generate exactly twelve vectors, just as six is the outcome in a plane. Had we started thus—with the choice of angles—we would have had to check the resulting vector lengths, to find out that indeed they are all the same. In either case, the end result is extremely satisfying.

VE: Results

Our first encounter with the vector equilibrium in Chapter 4—then we called it the "cuboctahedron"—illuminated its direct relationship to the octahedron and the cube. We now elaborate on our description of the VE, and before the end of this investigation we shall know almost everything about this extremely important shape. Let's begin here with its major characteristics.

Above all, it is the "omnidirectional arrangement of forces." This equivalence is unique to the VE.

Secondly, this shape bears an interesting relationship to other familiar polyhedra. Its twelve radii form eight symmetrically arrayed

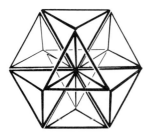

Fig. 7-7. Eight radiating tetrahedra alternate with six half octahedra.

regular tetrahedra—corresponding to the VE's eight triangular faces. Figure 7-7 emphasizes the tetrahedra, which radiate outward edge to edge, creating six cavities in the shape of square-based pyramids. Again, because of the uniform edge lengths everywhere, these cavities are actually perfect half octahedra, corresponding to the six square faces of the VE, which in turn correspond to the six faces of the cube, as was revealed by degenerate truncation in Chapter 4.

Thirdly, "the pattern of this nuclear equilibrium discloses four hexagonal planes symmetrically interacting and symmetrically arrayed... around the nuclear center" (981.11). If you look closely at Figure 7-8 the four hexagons are clearly visible: one parallel to the horizon, one in the plane of the page, and two more, slanted to the right and to the left, at 60 degrees to the horizon. As we might have expected, the vector equilibrium consists—in a way exclusively—of hexagons. The symmetrical properties of hexagons with respect to the plane are evident (refer back to Fig. 7-4), and so the discovery of intersecting hexagons in a spatial equilibrium of vectors is not surprising.

However, intuition cannot as easily predict the *number* of hexagons. An array of equivalent vectors (taking into consideration both magnitude and angular orientation) is achieved by exactly four evenly spaced intersecting hexagons. Thus the existence of four fundamental planar directions ("dimensions"?) describes one aspect of the inherent shape of space.

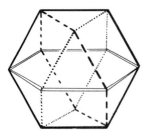

Fig. 7-8. Four hexagonal cross-sections of the VE.

These hexagons are exactly parallel to the four faces of the tetrahedron; having the same angular orientation, they are identical mathematical planes. The only difference is that in the VE they intersect at a common center, while in the minimum system they together enclose space.

Also fascinating is the fact that each of the twelve radiating vectors is perfectly aligned with an opposite vector—exactly 180 degrees apart. Thus the twelve can be seen as six intersecting lines with a positive and negative direction (each line twice the length of the original unit vector)—just as the XYZ axes are three lines intersecting to define six directions: three positive and three negative, evenly spaced with intervening angles of 90 degrees. Once again, these six intersecting lines are parallel to the tetrahedron's edges. It was not at all obvious from our initial requirements for a vector equilibrium display that the resulting radial lines would be collinear pairs, nor that these six (double-length) vectors would each lie in the same plane as two others, producing four precisely defined hexagons.

Our goal was to create a radial display of evenly spaced unit vectors. In so doing, we arrive at two fundamental observations about the order inherent in space: the existence of four distinct planes of symmetry and six linear elements. Both aspects are first exhibited in nature's choice of minimum system and secondly reinforced by her unique equilibrium configuration.

Degrees of Freedom

The subject of twelve fundamental directions of symmetry, with their six natural positive–negative pairs, leads directly to a discussion of the "twelve degrees of freedom" inherent in space. The term is almost self-descriptive, but can be best explained in reverse. That is, we explore the number of degrees of freedom inherent in space (and thus effecting every system) in terms of how many restraining forces are necessary to completely inhibit a system's motion. What is the minimum number of applied forces necessary to anchor a body in space?

Again, we can start with a planar analogy. Imagine a flat circular disk, such as a coaster, lying on a table and held in place by two taut strings pulling in opposite directions. The disk looks stable, but actually is free to move back and forth, at 90 degrees to the line of the two restraints (Fig. 7-9a). So, we try applying three tension

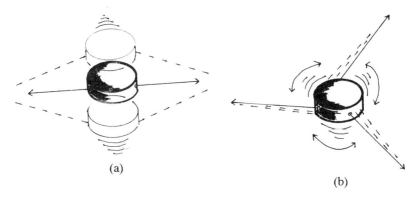

(a)

(b)

Fig. 7-9

forces, 120 degrees apart (Fig. 7-9a), and observe that the circle's position is fixed. Actually, it turns out that only the location of the exact center of the circle is fixed, for the disk is free to rotate slightly in place. Because rotation involves motion directed at 90 degrees to all three strings, there is nothing to restrain the circle from twisting back and forth, as shown in Figure 7-9b. Three additional strings to counteract each of the original restraints would have to be added to prevent all motion, for a total of three positive and three negative vectors.

In space, a similar procedure involves a bicycle wheel. Suppose that our goal is to anchor the hub with a minimum of spokes. At first glance this may appear to be the same problem as the previous planar example; however, in this case, the hub has both width and length. Both circular ends of the narrow hub—typically about a half inch wide and 3 inches long—must be stabilized. With only six spokes attaching the hub to the rim (three fixing the position of each end, as shown in Figure 7-10a), the system feels quite rigid; force can be applied to the hub from any direction—up, down, back or forth —without budging it. However, the hub has no resistance to an applied torque, the effect of which occurs at ninety degrees to the spokes, and is therefore able to twist slightly about its long axis. Three more spokes at each end, to counterbalance the original six, remove the remaining flexibility. All twelve degrees of freedom are finally accounted for, with a minimum of twelve spokes (Fig. 7-10b).

This experiment is quite rewarding to experience—well worth trying for yourself. You don't need to go as far as dismantling a bicycle wheel; just find a hoop of any material and size and a short

 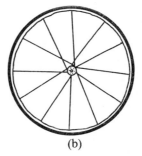

(a)   (b)

Fig. 7-10. Minimum of twelve spokes needed for stability.

dowel segment, and then connect the two with radial strings added one at a time until the hub suddenly becomes rigidly restrained. It is enormously satisfying to feel the hub become absolutely immobile (all "freedom" taken away), with the surprisingly low number of twelve spokes.[4]

What both the planar and spatial procedures indicate is that degrees of freedom are both positive and negative. In anchoring the hub of the bicycle wheel, there at first appear to be six degrees of freedom; however, each has a positive and negative direction. In conclusion, degrees of freedom measure the extent of a system's mobility: how many alternative directions of motion must be impeded before the body in space is completely restrained.[5]

The twelve vectors needed to restrain a body can also be omnidirectional, instead of the basically planar organization of the spoke wheel. Fuller takes us through a similar sequence in *Synergetics*, which starts with a ball attached to one string. The ball is free to swing around in every direction; the only restraint is on the radial "sweepout" distance. The ball's motion is thus free to describe a spherical domain. The addition of a second string restricts the ball to motion within a circular arc in a single plane (Fig. 7-11a). A third string allows the ball to swing only back and forth, in a linear path. The ball can always be pushed slightly out of place, no matter how taut the three strings (Fig. 7-11b). And just as, in our search for the minimum system, a fourth event suddenly created insideness and outsideness, by adding a fourth string to the ball, its position is suddenly fixed. ("Four-dimensionality" again.) In their most symmetrical array, the four strings go to the four vertices of an imaginary tetrahedron, and are therefore separated by approximately 109.47 degrees, the tetrahedron's central angle (Fig. 7-11c).

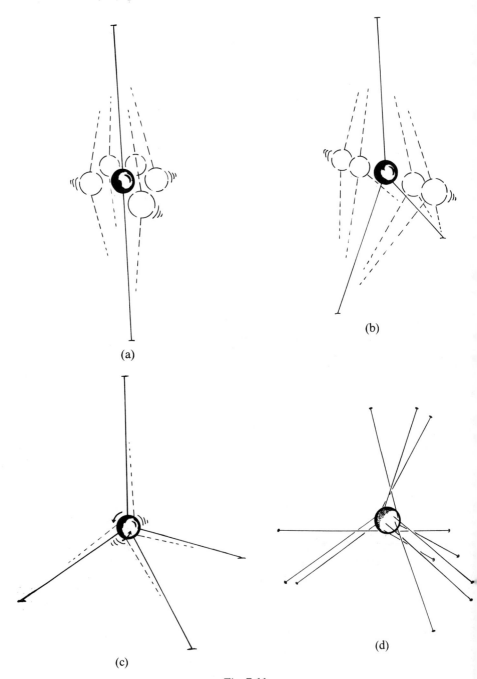

Fig. 7-11

But of course that's not the end of the story; the ball is still free to twist in place. To prevent this slight rotation, three strings must be attached to each location of the original four (Fig. 7-11d). This result is related to the fact that three is the minimum number of coordinates needed to specify the location of a point in space, with reference to the origin of a coordinate system.

The whole picture is falling into place. Every physical body has four basic sides, or four corners: two points alone are only collinear, and three are only coplanar; not until there are four corners can the property of spatial existence be recognized. As a result, any physical body must be held at four noncoplanar points, with three restraints at each point, in order to be stabilized (Fig. 7-11d).

This result suddenly ties in to the earlier discussion of pattern integrity. Triangles are necessary for stability. Therefore, while the ball was seemingly held in place by four restraints, it could still rotate locally because the locus of each individual restraint could not be stable without triangulation—subsequently provided by the addition of three strings per locus.

Bucky explains the situation further. Consider the ball with the original four restraints. The strings impinging on the ball create four vertices without supplying the necessary six edges to stabilize their position with respect to each other. They essentially form an unstable quadrilateral rather than a stable tetrahedron. The lesson is the same. There are four fundamental corners in every system, and each must be triangulated: $4 \times 3 = 12$. Thus there are twelve degrees of freedom, tetrahedrally organized. Twelve is a frequently recurring number in synergetics, a fundamental part of space and geometry, as we shall see again and again.

The above procedure describes Fuller's own interpretation of "degrees of freedom," which must be distinguished from Loeb's analysis of the concept, as briefly explained in Chapter 5.

The introduction to vector equilibrium is now complete except for one philosophical consideration. It is important to realize that the whole discussion is about conceptual—never actual—balance; equilibrium in any physical form can only be an approximation. No matter how exactly centered the hub of the bicycle wheel seems and no matter how tight the spokes, gravity's pull on the hub will always exert more tension on the upper spokes, leaving the lower spokes ever so slightly slack and imperceptibly curved. The balance is imperfect. Moreover, energetic motion never ceases. The air molecules in the living room do not stop their vigorous motion once the temperature is consistent. The floor and fallen object press together in a persistent dynamic exchange, encompassing furious activity at the atomic and molecular levels.

There is always motion in real systems: some (however minute) residual springiness in tension materials, as well as ever-present invisible bustling activity on the atomic scale. We cannot see the energetic motion in most systems, and so our perception is that of a state of perfect equilibrium. And indeed, for all practical purposes—that is, for a given level of resolution—we can have a stable balance.

But Fuller cannot in good conscience leave it at that. He reminds us that *real* equilibrium would mean an end to all, or "Universal death." An end to aberrations and imperfections is an end to motion and energy. All physical reality—life and nonlife alike—consists only of energy. Hence there is no absolute equilibrium:

> Nature is said to abhor an equilibrium as much as she abhors a perfect vacuum or a perfect anything. ... The asymmetric deviations and aberrations relative to equilibrium are inherent in the imperfection of a *limited* life.... Despite the untenability of equilibrium, it seemed to me that we could approach or employ it referentially.... A comprehensive energy system could employ the positive and negative pulsations and intertransformative tendencies of equilibrium. (420.041)

> The vector equilibrium is a condition in which nature never allows herself to tarry. The vector equilibrium itself is never found exactly symmetrical in nature's crystallography. Ever pulsive and impulsive, nature never pauses her cycling at equilibrium: she refuses to get caught irrecoverably at the zero phase of energy. (440.05)

All events, all systems exist as a result of their constant fluctuation in and out of ideal equilibrium—far too rapidly for perception. Fuller's goal was to develop a model for what that theoretical ideal must look like, in terms of spatial properties. But vector equilibrium is not a structure; he is quick to point out the distinction: it is a system—to be "comprehensively" grasped by "metaphysical minds":

> Synergetics... accommodates Heisenberg's indeterminism of mensuration inherent in the omniasymmetry of wavilinear physical pulsations in respect to the only metaphysical (ergo, physically unattainable) waveless exactitude of absolute equilibrium. It is only from the vantage of eternal exactitude that metaphysical mind intuitively discovers, comprehends, and equates the kinetic integrities of physical Universe's pulsative asymmetries. (211.00)

The concept of imperfection can only be held relative to the mind's grasp of theoretical perfection. In other words, "pulsative asymmetries" require a frame of reference in order to be defined and registered.

Time is responsible for these asymmetries. Separate time out of the picture, and you are left with the absolute perfection of time-lessness. Absolute equilibrium exists sub-time or meta-time; the passage of the shortest instant of time will reveal "pulsative asym-metries." But metaphysical mind has an all important need for timeless models, through which to understand Universe.

8

Tales Told by the Spheres: Closest Packing

Much has been written over the years by mathematicians and scientists about the problem of "closepacking" equiradius spheres. It's not a subject that the rest of humanity has tended to get excited about; however, the orderly patterns revealed by these packings are unexpectedly fascinating. Closepacking equiradius spheres might at first sound like the type of abstract mathematical game Fuller railed against; after all, there's no such thing as a sphere. But if nature exhibits no examples of pure spheres—that is, no perfectly continuous surfaces equidistant from one center—we can still discuss the concept of a spherical domain. Imagine various approximations of the model, such as a soap bubble or, less fragile, a Ping Pong ball. The concept of multiple equiradius spheres turns out to be quite useful, providing a superb tool with which to investigate the properties of space. Let's look into some of the reasons why.

Equilibrium: Equalization of Distances

The connection to equilibrium is perhaps the most important reason to experiment with sphere packing. A sphere is defined as the locus of all points at a given distance from a central point; consequently, in an array of tangent spheres, their centers will be separated by a uniform distance. The configuration developed in the previous chapter to represent vector equilibrium—requiring equal lengths in all directions—can be created quite simply with the aid of this model. If one sphere is completely surrounded by a number of spheres of the same size, the distances between the internal sphere's center and the centers of all surrounding spheres are necessarily the same as the distance between the centers of adjacent external spheres, provided all spheres are in contact with each other. The resulting cluster is shown in Figure 8-1, along with a cross-section of the packing to illustrate that the distances are the same. Closepacked spheres automatically set up an array of evenly spaced points. Equal distances represent balanced forces: ergo, equilibrium.

Fig. 8-1

Symmetry versus Specificity of Form

A sphere is the form of "omnisymmetry" in spatial reality. Symmetry describes the degree to which a system can be rearranged without detectable change. The sphere's shape presents no corners, no angles —in short, no landmarks—by which to detect rotation or reflection. Its very shapelessness enables us to explore the shape of space. Furthermore, the total absence of angular form makes the precisely sculpted shapes generated by packing the identical "shapeless" units together all the more surprising. It is easy to see that individual spheres, as omnisymmetrical forms with neither surface angles nor specific facets to mold the form of clusters, cannot determine through their own shape the overall shape of packings. In conclusion, we are not so much interested in the ("nondemonstrable") spheres themselves, as in using sphere-packing as a medium through which spatial constraints can take visible shape.

Organization of Identical Units

Finally, the standard model of an atom is spherical: packets of energy are spinning so rapidly about a tiny nucleus that the atom can be considered occupying a spherical domain. (In fact, the orbit of any object spinning in all directions defines a sphere.) We can therefore pack spheres together in the hope of learning about atomic and molecular aggregations. To state the problem more generally, the organization of identical units is an important theme in biology and chemistry. All sorts of units—such as atoms, molecules, cells, DNA nucleotides—must be organized to function cooperatively in structures far more complex than the individual units themselves. Spatial constraints are responsible for much of the superb organization of biological phenomena—allowing and encouraging certain configurations while prohibiting others. Yet, despite the influential role of space, scientific thought does not as a matter of course take this into consideration.

Sphere-packing can be thought of as a method of blindly gathering evidence; we experiment with these identical units without knowing the outcomes, and space enters in to direct traffic. The resulting configurations are absolutely reliable. We are thereby able to observe the shape of space, manifesting itself through the innocent spheres.

New Level of Focus

Despite our discipline of viewing whole systems, we have reached a point at which we must zoom in to look closely at certain details of Fuller's *Synergetics*. The sections called "Closest Packing of Spheres" contain some of the most difficult passages in his book, rendering Fuller's observations inaccessible without considerable perseverance. Not only is the description hard to follow, but these patterns seem to elude application. It is therefore especially important to understand the logic behind Fuller's use of sphere-packing in an investigation of nature's coordinate system. Otherwise, it will be difficult to see how these details fit back into the big picture. Even though the immediate goal of this text is to clarify the configurations described by Fuller, a list of results, no matter how clear, is not likely to be interesting unless the premise behind the search is understood. At this point, the reader may even have thought of further reasons to add to the ones stated above, for there are many dimensions of this issue. However, as it probably remains difficult to predict or visualize the patterns themselves, we bravely proceed.

Background: Closepacking

Packing spheres together with a minimum of interstitial space is a problem that still presents a challenge to mathematicians.[1] (Actually the problem has been solved, but it turns out to be extremely difficult to prove that the solution is indeed maximally dense.) For our goal of exploring Fuller's studies, we only deal with one of the two types of closest packing described below.

Once again, we start with the plane in order to establish a firm hold on the concept. Suppose we want to fit the largest possible

Fig. 8-2

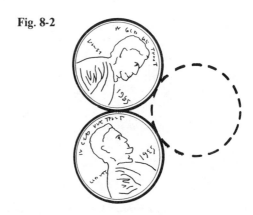

number of pennies on the surface of a small table—another way of saying we want the pennies to lie as close to each other as possible. As we saw in the previous chapter, a square grid of tangent pennies wastes considerably more table space than a triangular array. Observe in Figure 8-2 that two tangent pennies create a valley that naturally embraces a third penny. All pennies are therefore allowed as close together as physically possible if every penny is situated in a valley created by two others.

In the same way, there is only one closest-packing arrangement of spheres in the plane: each sphere must be in contact with six others (Fig. 8-3). A second identical layer can be placed on top of the first, with its spheres all landing in nests created by three neighbors on the

Fig. 8-3. Six spheres closepack around one.

first layer. To achieve our goal of packing spheres as close together as possible, we have thus far had no choice as to the next step. A third layer however can be superimposed on the second in one of two different ways to maintain maximum density. The spheres of the third layer can either be placed directly above the *spheres* of the first layer or above the *nests* in the first layer. A schematic comparison of the two packings is shown in Figure 8-4. The former is called

Fig. 8-4. Cubic packing (top) versus hexagonal packing (bottom), showing three layers of spheres in each packing.

hexagonal closepacking; the latter, cubic closepacking. In both cases, every sphere touches exactly twelve others—as we might have anticipated from our VE studies. The difference between these two packings is explained in the following description.

Instead of trying to imagine indefinitely large planar expanses, we focus on a small portion of the closepacking—the arrangement surrounding a single sphere. We start with one ball on a table; six others closepack around the first, and find themselves exactly tangent to each other. As with the pennies, there is no choice as to the number of spheres in the cluster. The flat hexagon of spheres creates six separate "nests" of three spheres each, as seen in Figure 8-3, but additional spheres of the same size, sitting in any one of the six

(a)   (b)

Fig. 8-5. Hexagonal (left) versus cubic (right): twelve spheres pack tightly around one.

nests, partially block adjacent nests. As a result, there is only room for a ball in every other nest, allowing a total of three nesting spheres on the hexagonal cluster. These three balls (by magic or else by inherent spatial properties) rest exactly tangent to each other, a perfect equilateral triangle (Fig. 8-5a, top). The arrangement has neither leftover space nor crowding; all three spheres on top are tightly shoved into nests, and—because the planar group (six around one) are clearly as close together as possible—all ten spheres are convincingly closest-packed.

We can then flip the whole package over and repeat the procedure on the other side. We again have two choices: the second three-ball addition can be placed either directly above the three on the bottom (meaning both the top and bottom triangles are pointing the same way), or it can be oriented the opposite way (Fig. 8-5a, b). The former choice (hexagonal) outlines the vertices of a polyhedron in which the squares are adjacent to other squares (in three pairs meeting at the "equator"), as are six of the eight triangles (Fig. 8-5a). The latter choice (cubic) consistently alternates triangles and squares, so that squares are entirely framed by neighboring triangles—and triangles by squares (Fig. 8-5b).

The latter arrangement—"our friend the vector equilibrium" as Bucky says—is the more symmetrical of the two choices, and is therefore used as the basis of Fuller's subsequent sphere-packing studies.[2]

Planes of Symmetry

The following observations pertain to an indefinite expanse of cubically closepacked spheres. For example, imagine a room full of Ping Pong balls so tightly nested together that every ball touches exactly twelve others as described above. The idea of a room full of balls at first suggests such chaos that the precise organization arising out of the requirements of closest packing is truly remarkable. The array contains seven planes of symmetry, characterized by two different types of cross-section. The first is obvious, because we generated the cubic packing with successive layers of triangulated planes. However, it is not necessarily obvious that there are four different orientations of triangulated planes—parallel to the four faces of the tetrahedron (and therefore to the VE's four intersecting hexagons). Even though we built this array by stacking triangular layers of balls in only one direction, the result incorporates parallel triangulated layers in four different directions. (We further note that all four hexagons of the

VE are preserved in cubic packing, whereas in hexagonal packing, hexagons are formed in only one orientation, the horizontal plane.)

Secondly, there are three distinct planes characterized by a square pattern of spheres. This discovery seems to contradict our expectations, for we have learned that squares are not closepacked. However, the emergence of three mutually perpendicular square patterns is an inescapable by-product of nesting triangular layers. The square planes correspond to the VE's square faces, which consist of three mutually perpendicular pairs of parallel faces—just like the faces of a cube.

In a space-filling array of closepacked spheres, these seven planes extend indefinitely—with neither curve nor bend. A cross-section of such a packing has a square or triangular arrangement, depending on which way the packing is sliced. So although we started with only triangulated layers (in order to create a maximally dense array of spheres), square cross-sections automatically arose, just as octahedral cavities automatically arose next to the radiating tetrahedra in the vector equilibrium. This is the shape of space.

It is interesting to note that, although when we stacked triangulated layers it was necessary to make a decision at the third layer that led to two different packings, if we were to start out instead with the square layers (unstable though they may be), there is only one way to proceed. Successive square layers can be placed on top of each other so that each ball lands in a square nest (as opposed to being placed directly on top of another ball, creating an array of cubes which would clearly not be closepacked). The internesting of layers stabilizes the otherwise unstable separate layers. Once two layers are packed together, every ball nests in a group of four balls on the adjacent layer—creating half-octahedral pyramids (Fig. 8-6) separated by the inevitable by-product tetrahedra. Continuing to stack square layers in this way, cubic packing—rather than hexagonal—emerges. The vector equilibrium array is thus generated automatically, with no decisions along the way, by simply stacking square layers. Each and every ball is surrounded by exactly twelve others, in the more symmetrical of the two possibilities.

It is satisfying to reflect on the exquisite logic of this tradeoff: although balls arranged in square patterns are not as closely packed as triangular planes, the nests are deeper. A ball placed in any four-ball nest (to start a second layer) sinks deeply into the cluster; in comparison it seems perched on top of the tight triangular nest. Therefore, successive square layers, although inefficient in themselves, fit more closely together than triangulated layers.

Fig. 8-6. Five oranges creating half-octahedron.

Part of the challenge to mathematicians in proving that the hexagonal and cubic packings qualify as the solution to minimizing interstitial space is the fact that spheres in these two packings occupy just over 74% of the available space, while the four-ball tetrahedron alone is able to occupy 77.96% of its overall volume.[1] It is easy to accept that four balls cannot be pushed any closer together than the tetrahedral cluster and accordingly that the latter figure is the maximum density. Therefore, 74% seems insufficient—not easy to accept as *the* solution to the problem of closest packing. However, there is no getting around the fact that the six-ball octahedral cluster (Fig. 8-7) is less dense than the four-ball tetrahedron—as the former has more leftover room in the middle—and that the constraints of space are such that tetrahedral sphere clusters simply cannot be extended indefinitely by themselves. Attempting to fill space exclusively with tetrahedral groups, we quickly discover awkward leftover gaps—spaces not quite big enough to contain another sphere. In order for spheres to be both consistently tangent and tightly nested together, we have to allow the naturally alternating tetrahedral and octahedral clusters. The problem would be remarkably easy if spheres

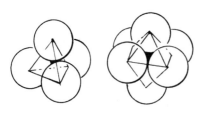

Fig. 8-7. Tetrahedron and octahedron.

could pack tetrahedrally in an indefinite array, but they cannot. No amount of force can change this constraint; space is simply not shaped that way. Twelve around one, creating 60-degree angles both radially and axially, with alternating tetrahedra and octahedra, is the closest packing.

Fuller Observations

Having settled on the most symmetrical and dense sphere packing, to faithfully present the characteristics of space, we are ready to explore the shapes and "periodicities" observed by Fuller. The reliable precision of these patterns indicates that they are molded by space, not by imposed design.

What are these patterns that captured Fuller's attention so long ago (and kept it for fifty years)?

We start very simply with the phenomenon of "triangular numbers," a sequence of numbers in which each successive term is equal to the previous term plus the number of terms so far. These numbers can be generated by triangular collections of balls, arranged as in a rack of billiard balls. The total number of balls in each triangle, in a series of progressively larger groups, is a triangular number. Figure 8-8 shows the first five groups. The first member of the sequence is "1"; the second is obtained by adding two to the first, to get "3"; the third, by adding three to the other two, to get "6"; and so on. Each successive number is generated by the addition of a row with one more ball than the last. The sequence of numbers thus generated $(1, 3, 6, 10, 15, 21, 28, \dots)$ is specified by $(n^2 - n)/2$ for $n = 2, 3, 4, \dots$. ($n = 1$ corresponds to 0, which is not—strictly speaking—modeled by a triangle).

1   2   3   4   5

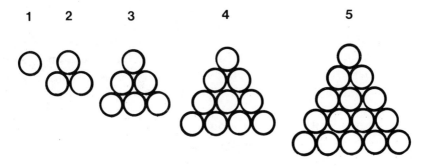

Fig. 8-8. Triangular numbers.

The formula $(n^2 - n)/2$ might appear to be a more difficult way to obtain these values (and certainly for small groups its easier just to count balls), but of course for the twentieth or even the ninth triangular number, it's far more direct to subtract 10 from 10^2 and divide that by 2 to get 45 than to draw rows and rows (nine rows) of circles! Mathematics presents a shortcut—otherwise known as a generalized principle.

So far we have a numerical progression $(n^2 - n)/2$ that happens to be modeled by triangular clusters. We might have chosen to discuss a variety of other sequences, for example, the numbers generated by n^2: $1, 4, 9, 16, 25, \ldots$, which could be labeled "square numbers" because they are geometrically represented by square clusters. But we have a specific motivation for paying attention to triangular numbers, because of one especially significant characteristic: the nth term is the number of relationships between n items, for $n = 1, 2, 3, 4, \ldots$. For example suppose that for any given number of people, we wish to know how many telephone lines are required to link everyone to everyone else by a private line. The answer is the triangular number corresponding to a number of rows one less than the number of people. Two people require only one line; three require three; four require six (as portrayed in Figure 3-1), and n require $(n^2 - n)/2$ private lines. It is no longer far-fetched to imagine applications for this formula. The real lesson from triangular numbers is that significant algebraic expressions, such as $(n^2 - n)/2$, the number of relationships between n events, can be represented geometrically. That much established, we proceed to the next development.

Tetrahedra

We can stack triangular arrays of decreasing size, creating tetrahedral clusters. (We could thereby isolate a sequence of "tetrahedral numbers.") But let's go back to the very beginning.

One sphere alone is completely free to move, and closepacking is of course not an issue. Suppose we have two billiard balls tangent to each other. If the only requirement is that the two balls stay in contact, we shall observe that they are free to roll around each other's entire surfaces (Fig. 8-9a). We then introduce a third ball, allowing it to roll in any direction while touching at least one of the first two balls. It eventually rolls into the valley between the other two, establishing a naturally stable triangle (Fig. 8-9b). At this point we require all three balls to stay in contact, and discover that they

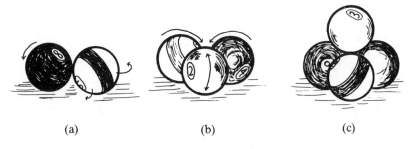

(a)   (b)   (c)

Fig. 8-9. "Four balls lock."

are still able to roll, but only inward or outward (toward or away from the triangle's center, in tandem) "like a rubber doughnut," to use Bucky's words. The freedom of motion of each sphere is thus considerably more limited—spinning about one specific horizontal axis instead of unrestrained motion in every direction.

A fourth ball rolls across the surfaces and lands comfortably in the triangular nest. Suddenly, all four balls are locked into place, unable to roll or move in any direction (Fig. 8-9c). This is the first stable arrangement, with the requisite minimum of four. The tetrahedron is once again at the root of our investigation.

At this point, it may be illuminating to construct some of these structures, for example with Styrofoam balls and toothpicks, or small plastic beads and glue. One particularly satisfying demonstration involves bringing four spheres together and trying to create a square. It is easy to feel how unstable that arrangement is: the balls gravitate naturally toward the tight tetrahedral cluster. Fuller placed considerable emphasis on the benefits of hands-on construction to gain thorough familiarity.

The theme of Fuller's tetrahedral sphere packings is the presence or absence of nuclei. The word "nucleus" evokes the image of a central ball spatially surrounded by other balls, which is exactly the way Fuller uses it for the VE packings. However, his observations about tetrahedral patterns are based on a somewhat different approach. Most of the inpenetrability of the sphere-packing sections in *Synergetics* can be removed with one simple clarification: a "nucleus" in VE packings is defined as a ball at the geometrical center of the whole cluster, whereas a "nucleus" in tetrahedral stacks is a ball at the exact center of an individual planar layer.

Start with the four-ball tetrahedron developed above, which consists of a fourth ball added to a triangle of three others. Next, the simple four-ball tetrahedron is placed on top of a flat six-ball

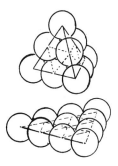

Fig. 8-10

triangular base, creating a tetrahedron with three balls per edge (Fig. 8-10). There are three layers, with ten balls altogether— six plus three plus one. Throughout his sphere-packing studies, Fuller uses the number of tangency points per edge (in other words, the number of *spaces between* spheres along an edge of the cluster, rather than the number of spheres) for the assignment of frequency. The four-ball tetrahedron is thus "one-frequency" (as is appropriate for the first case), and the next case, the ten-ball tetrahedron, is "two-frequency." We can visualize that each sphere-cluster polyhedron corresponds to a line drawing (or toothpick structure) in which the spheres' centers locate vertices which are interconnected by lines (or toothpick edges) through tangency points. Recalling that "frequency" is defined as the number of modular subdivisions, the justification for Fuller's frequency assignment is evident from this translation, because the number of *subdivisions* (or line segments) per edge corresponds to the number of *spaces* between spheres, rather than to the spheres themselves, which correspond directly to the vertices (Fig. 8-10).

Triangular clusters, each with one more row than the last, are stacked to create larger and larger tetrahedral packings. We place the ten-ball (two-frequency) tetrahedron on top of the next triangular base, which itself consists of ten balls, to get a three-frequency tetrahedron, with twenty spheres altogether (Fig. 8-10). We have thus begun a list of values that might be called *tetrahedral* numbers: 4, 10, 20, followed by 20 plus the additional triangular layer of 15, to total 35 (Fig. 8-11). The progression can be continued indefinitely.

Consider the different individual layers. There is a *ball* in the exact geometric center of some of the triangular groups, while others, having three balls around the exact center instead, are left with a *central nest*. Successive triangular clusters reveal a specific pattern: every third layer has a central ball, or nucleus. Fuller describes this progression as a "yes-no-no-yes-no-no" pattern. To see how it works, let's look at the first few members of the sequence. One ball

Fig. 8-11

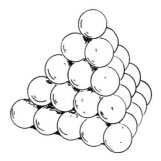

alone is automatically central—"a potential nucleus" in Fuller's words; it earns a "yes." The next layer, the three-ball triangle, has a nest, but no nucleus; that merits a "no;" likewise for six ("no"). Not until the ten-ball (three-frequency) group is there a nucleus—shown as the dark ball in Figure 8-12 ("yes" again). Each "yes" case (with nucleus) consists of a hexagonal arrangement with three additional corners tacked on, to complete a triangle. The rest ("no" cases) are organized triangularly from the center out to the corners—simple threefold rotational symmetry, containing no central hexagon.

Fuller calls our attention to this periodicity (or pattern) within the system: starting from the top, the pattern is Y-N-N-Y-N-N-Y...

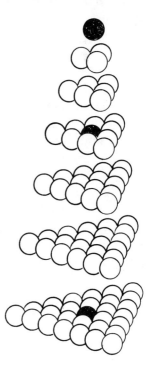

Fig. 8-12. "Yes-no-no-yes-no-no."

(notice that "N" can stand for "nest" as well as for "no"):

> 415.55 Nucleus and Nestable Configurations in Tetrahedra: In any number of successive planar layers of tetrahedrally organized sphere packings, every third triangular layer has a sphere at its centroid (nucleus.)

Fuller's yes-no-no pattern describes the presence of nuclei in certain layers of a pyramid; he does not ask which pyramids of gradually increasing frequency contain an overall nucleus (at the center of gravity.) This is a subject open for further exploration, which we leave for the time being as we continue to explore Fuller's material.

Vector Equilibrium

In cubic packing, twelve spheres surround one sphere, with each sphere tangent to every neighbor and without any gaps. This perfect geometric fit of the thirteen omnisymmetrical forms provides a basis for understanding the fundamental directions inherent in space. But we saw that the configuration is more specific than just a numerical consistency; the spheres outline the vertices of the cuboctahedron, or VE. This shape seems to appear out of nowhere. Created by the cluster of cornerless spheres shoved together, this result is as counter-intuitive as it is reliable. The more we learn about the shape of space, however, the more natural the appearance of the VE—or any manifestation of "twelve degrees of freedom"—becomes.

Frequency

The six squares and eight triangles outlined by the closepacked spheres, although unmistakable even in the simplest case, become more and more distinct as the frequency increases. As Fuller's convention is to refer to the number of spaces (rather than spheres) along the "edge" of the cuboctahedral cluster as the frequency of the system, the first twelve-around-one group is "one-frequency." The next layer—in the VE case, a surrounding envelope rather than just another layer added to the bottom—is two-frequency, having three balls along each edge. The next is three-frequency, with four balls per edge, and so on. It follows that the higher the frequency, the smaller an individual sphere is in relation to its polyhedral face, and so these polygonal faces look progressively less bumpy, or more sharply defined (Fig. 8-13). The precise planar organization of the clustered spheres becomes more obvious as the frequency increases.

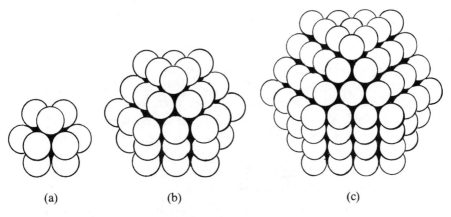

(a)   (b)   (c)

Fig. 8-13

The appearance of eight triangles and six squares was not an accidental property of the first layer: the VE is here to stay. This lesson is continually reinforced by additional layers.

Having established the *shape* of symmetrical nuclear sphere-packing we proceed to investigate *numbers*. Twelve balls fit tightly around one; how many does it take to completely surround the twelve with a second layer? By carefully placing balls in the "nests" on the cluster's surface, we generate a two-frequency VE shell of exactly forty-two balls. We might begin by placing one ball at the center of each of the six squares and then in each nest along the twenty-four VE "edges", thereby superimposing six two-frequency squares (nine balls each) over the simple four-ball squares (Fig. 8-13b). The edges of the six two-frequency squares supply the spheres for the adjacent two-frequency triangles (without adding any more balls). These second-layer triangles cover the eight three-ball triangles of the first layer. The twelve vertices, or corner spheres, are each shared by two squares, and so twelve must be subtracted from fifty-four (nine spheres per square times six squares) to get the total number for the second VE layer: forty-two spheres. (See Fig. 8-13.)

Envelop the whole package with a third layer, a three-frequency shell with four balls per edge. Following the above procedure, we count sixteen balls in each of the six squares, for a total of $16 \times 6 = 96$, from which we must subtract the twelve vertices counted twice due to overlap between squares. The edges of the eight triangles are again already in place—provided by the edges of the squares—but on this shell, a central sphere which belongs in each three-frequency triangle must still be added; eight more spheres are therefore needed to complete the enveloping layer. 96 minus 12 plus 8 yields a total of 92 balls.

The next shell (four-frequency) requires 162 balls; the five-frequency layer consists of 252, six of 362, and so on. We are now able to detect a pattern by looking carefully at these numbers: $12, 42, 92, 162, 252, 362, \ldots$. It will come as no surprise to the observant student of numbers to learn that the next shell consists of 492 balls.

What exactly is going on? To begin with, we notice the consistent last digit: every single number ends with 2, reminiscent of Euler's law and its "constant 2." Fuller interprets this persistent "excess of 2" in radial sphere packing as further affirmation of the inevitable "poles of spinnability," inherent in the topology of singly closed systems. And indeed, the temptation to embrace a single explanation is strong. The fact that the number of spheres per shell always ends with the digit 2—even though those numbers increase drastically with each successive layer—seems too strange to ignore; we want an explanation for nature's behavior. But at this stage, speculation as to significance is a sidetrack: our task is to fully describe the configurations. As soon as we fully understand the patterns and are thus armed with the facts, such speculation will be appropriate and indeed inevitable.

After observing the reliable last digit, we can simplify our sequence—following Fuller's procedure—by subtracting the 2 from each term, removing the distraction to assist further analysis. We are left with $10, 40, 90, 160, 250, 360, 490, \ldots$, all divisible by 10. So let's divide by 10. This leaves $1, 4, 9, 16, 25, 36, 49, \ldots$, and now the pattern is clear.

The latter sequence is generated by f^2, for $f = 1, 2, 3, 4, 5, 6, 7, \ldots$. We choose "f" in this case, to represent frequency, for it turns out that the relationship between frequency and number of units per shell can be directly specified. Nature thus reveals yet another "generalized principle." This equation actually describes a straightforward edge-length-to-surface-area relationship, exactly what we expect from geometry—in a slightly different format.

The next question is how to specify the relationship in precise terms. We work in reverse from our final sequence ($f = 1, 2, 3, 4, \ldots$) to generate the original sequence. First we must raise the frequency f to the second power, then multiply each term by 10, and then add 2: $10f^2 + 2$ therefore gives us the total number of spheres for any shell (specified by frequency) in nuclear sphere packings.

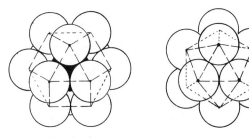

Fig. 8-14. Removal of nuclear sphere.

Icosahedron

We return to the initial twelve-around-one cluster, to try a new twist (literally). Imagine that we have thirteen spherical balloons—instead of Ping Pong balls—so that we can we reach through with a long pin to puncture the nuclear balloon, causing it to slowly deflate. As the nuclear balloon disappears, the twelve outside balloons shift, closing in symmetrically toward the empty space in the center. They can't go far—only enough for the six unstable squares of the VE to "tighten up" into two noncoplanar (or hinged) triangles. Squares were stable in the VE only because the nuclear sphere held them in place; the array was thus stabilized by triangulation in radial directions, producing alternating tetrahedra and half octahedra. Now, without the nucleus, the surface must be stable by itself; it therefore must be triangulated (Fig. 8-14).

The VE configuration has twelve "vertex" spheres. What shape do these twelve spheres become when fully triangulated? The six squares transform into two triangles each, which (added to the original eight) make twenty triangles in all. Twenty triangles and twelve vertices? That is none other than our friend the icosahedron, largest of the regular polyhedra, one of the "three prime structural systems in Universe."

We can create higher-frequency versions of the icosahedron, but they will always be single-layer shells. The icosahedral configuration arose as a result of removing the VE nucleus; the remaining spheres move in, partially filling the gap, and thus their positions no longer allow a space-filling array of spheres. Because icosahedral clusters are completely triangulated, they cannot be extended either inwardly or outwardly; they lack the necessary alternation of tetrahedra and octahedra.

Such omni-triangulated sphere-packing shells can have any frequency, despite being restricted to single-layer construction. Let's see what happens. Consecutive higher-frequency icosahedral shells cannot surround a previous layer as they do in the VE—which of course *starts* with a nucleus and continually surrounds it with layers. Icosahedral shells simply do not nest together. Instead, progressively larger, or higher-frequency icosahedra must be built one by one, each with one more sphere per edge, and always single-thickness.

What will happen to the relationship between frequency and number of spheres on a given shell? It turns out that—while both shape and volume change considerably—the *number* is unaffected by this transition from the VE's fourteen faces to the twenty icosahedral triangles. We verify this fact through the following observations. Notice in Figure 8-15 that a square pattern of spheres can be compressed into a rhomboid (diamond) shape without changing the number of spheres. The resulting diamond is more tightly packed than the square and consists of two triangles of the same frequency as the original square, sharing one edge, that is, the row of spheres that used to form the diagonal of the square. Figure 8-15 shows how the spheres of two triangles on an icosahedral shell correspond to one square face of a VE of the same frequency. And therefore, because the spheres on an icosahedral shell are *all* as closely packed as possible (as opposed to the VE, which alternates triangles with the more loosely packed square faces), a smaller, denser shell is produced, with the same number of spheres as a VE shell of the same frequency. $10f^2 + 2$ therefore also applies to icosahedra.

The icosahedron contains as much interior volume relative to surface area as is possible with only one type of face. Ever economical, nature therefore chooses icosahedral symmetry for the construction of a shell made of identical units; requiring a minimum of effort, this arrangement can arise automatically. Maximum volume, minimum material. It is thus easy to account for the icosahedral symmetry detected in the isometric virus capsid, the tough protein shell created by nature to house and protect the more fragile genetic material within, which is the source of the virus's instructions.[3] Nature consistently exhibits elegant solutions to design problems, because she finds the most efficient, or least energetic, way to operate. She has no choice but to adhere to the constraints of space. The example of the spherical virus shell is worth our brief attention, for it provides an elegant illustration of the "design science" of nature at work.

Let's examine the criteria: (1) A container must be constructed out of a large number of identical constituents (protein molecules), (2)

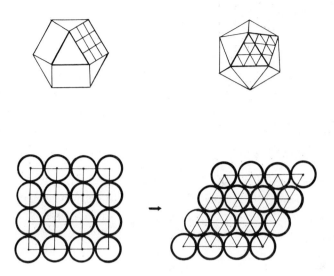

Fig. 8-15

for reasons which will be explained below, the shell must be able to self-assemble—that is, build and rebuild itself automatically, (3) maximum symmetry is advantageous to minimize the energy required for attractive bonds between the capsid molecules, and (4) the arrangement must be stable, which means triangulated.

The elegance of the relationship between structure and function is well documented in modern biology, and the isometric virus is no exception; its structure must be suited to its specific functions. A tough shell is to completely enclose minute amounts of genetic material—quantities necessarily insufficient for carrying detailed bonding instructions—yet it must easily disassemble and reassemble itself in order to release the viral genetic material into a host cell. The overall structure must therefore be dictated by properties of the subunits and by the constraints of space itself—both criteria also establishing a built-in check system.

A sphere, which maximizes the volume-to-surface-area ratio, is the key to an efficient solution. Interconnected molecules, which can only approximate that theoretical sphere, will achieve a spherical distribution most efficiently through icosahedral symmetry. Observations of isometric viruses have consistently revealed icosahedral patterns, thus reconfirming that nature chooses optimal designs.[4] We shall study this configuration in more detail in Chapter 15.

Further Discoveries: Nests

Throughout our investigation, we note the recurrence of a limited inventory of polyhedral shapes. This is perhaps revealed most dramatically by the behavior of closepacked spheres. A striking example is found in the results of an extra sphere placed in the central nest of flat triangular clusters. The first case is already quite familiar: a sphere placed in the minimum triangle of three spheres produces a regular tetrahedron, assuming all four spheres are the same size. A half octahedron is born out of a sphere nesting in a square group of four spheres, and we might reproduce any number of familiar shapes by putting equiradius balls together, but here we confine ourselves to the triangular clusters, for the scientific method tells us that the strength of an experiment often rests in drawing boundaries. Results thus obtained may lead to broader generalizations about related questions.

Remember our observation that every "N," in Fuller's Y-N-N pattern, represents a layer with a nest, since any triangular cluster without a central ball has a central nest. The center of a ball added to the first nest becomes the fourth vertex of a regular tetrahedron; the next group, six balls, also has a central nest, so a seventh ball is put in the space. Magically, the shape thus created is a significant one: the semisymmetrical tetrahedral pyramid, which is exactly one-eighth of the regular octahedron (Fig. 8-16). To understand what is meant by "one-eighth" of an octahedron, imagine that we slice a regular octahedron made out of "firm cheese," to use one of Bucky's images, in half—creating two square-based pyramids (Egyptian style). Then, cut both halves into quarters, to get eight octants, each with an equilateral triangle base and three isosceles-triangle side faces (45°-45°-90°). The octahedral central angle is 90 degrees, a fact which will prove especially significant in the next chapter.

The seventh ball is simply placed in the nest of the six-ball triangular layer to complete the four vertices of an "octant." Nothing in our sphere-packing investigation thus far would lead us to predict the appearance of this important shape, which is already a significant part of our polyhedral inventory. (Such unpredictability is getting to be a pattern.) So we proceed to the next case.

The ten-ball triangle has a nucleus (1: yes; 3: no; 6: no; 10: yes; ...) and therefore no nest. So we skip to the fifteen-ball cluster and, as before, drop a sixteenth ball in the central nest. We are no longer surprised to discover that the resulting shallow pyramid is

Fig. 8-16

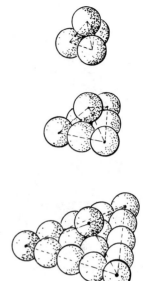

also a very special shape: one-quarter of a regular tetrahedron—a portion encompassing the volume from the tetrahedron's center of gravity out to any one of its four faces (Fig. 8-16). The tetrahedron's central angle, 109.471 degrees (109°28′16″), seems so irregular that the sense of coincidence is underlined.

With fifteen balls in the plane and a sixteenth in the center, this pyramid is quite shallow—and in fact, as a section of the minimum system, it is Fuller's terminal case. For each of the first five triangular numbers without nuclei, a sphere placed in the central nest forms an important shape in the VE–octet framework, an idea that we shall explore in greater detail in the next chapter. We thus have come to the end of this particular experiment, with the conclusion that closepacked spheres automatically yield many significant geometric shapes.

"Interprecessing"

The essence of precession, to Fuller, is 90°. And indeed, the counter-intuitive or mysterious thing about the behavior of gyroscopes (and other examples of precession in physics) is the resultant motion in a direction ninety degrees away from that of an applied force. For example if a downward force is imposed at the north point of a

Fig. 8-17

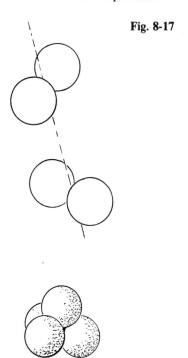

gyroscope spinning clockwise, it will tilt toward the east: 90 degrees away from the direction one intuitively expects. Fuller's "interprecessing" involves two systems "precessing" together, which he uses to mean oriented at 90 degrees with respect to each other. In his sphere-packing studies, "interprecessing" reveals subtle facets of symmetry which might otherwise go unnoticed.

We start with the simplest case: two identical pairs of tangent spheres, parallel to each other and separated by some distance. Rotate one of the two-ball sets 90 degrees, and then move the two pairs toward each other until they meet in the middle, so that the midpoints, or tangency points, of the two pairs are as close together as possible. (Notice that without the 90-degree twist, the result of bringing the parallel pairs together would be a square—unstable and not closepacked.) What is the result of this simplest case of interprecessing? A tetrahedron, of course (Fig. 8-17).

In retrospect, the answer appears obvious, for the initial condition of four spheres—the necessary ingredients of a tetrahedron—gives it away. However, the experiment highlights the 90-degree symmetry of the tetrahedron, which is otherwise obscured by the predominance of triangles and 60-degree angles. Rather than elaborating on the tetrahedron's right-angle symmetry here, we shall allow subsequent

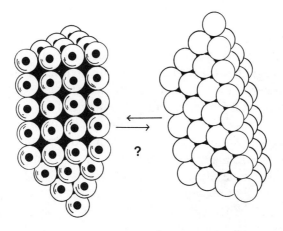

Fig. 8-18a

demonstrations to further illustrate this orthogonal characteristic. (See especially Chapters 9 and 10.)

Take two identical sets of sixty spheres, closepacked as shown in Figure 8-18a. Their irregular trapezoidal shape eludes immediate identification. That they do not seem to be a part of our familiar group of shapes is confirmed by numerous experiments in which participants are given these two pieces and asked to put them together in some way that seems correct. Countless false moves involve bringing similar faces directly toward each other, and again and again, the identical halves are put together in unsatisfying and incorrect ways. The correct solution is rarely discovered by the uninitiated—but once seen is unmistakable. This problem (simple, after the fact) is initially challenging because it is so hard to get

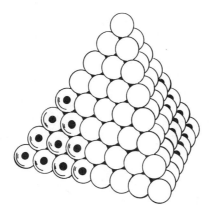

Fig. 8-18b

beyond the natural assumption that the two halves must approach each other directly—as if one half were approaching its own reflection in a mirror. What actually has to occur of course is that one half rotates 90 degrees with respect to the other (interprecessing) and the two rectangles mesh together perfectly, at right angles. "Wow!" Bucky would exclaim, apparently as surprised as his audience. The surprise is genuine in a sense, for the result is visually striking even if one already knows the answer: a perfect tetrahedron, eight balls per edge, or seven-frequency (Fig. 8-18b.)

Along the same lines, we now look at 60-degree twists. An especially pleasing example involves two simplest triangles, of three spheres each. The triangles face each other directly; then one rotates 60 degrees before pushing them together, and the result is an octahedron. The six spheres are precisely situated as octahedron vertices, framing eight triangles. (Refer back to Fig. 8-7.)

Next, let's take two eighth-octahedron seven-ball sets (the six-ball triangle with a seventh ball in the central nest). The two triangular bases of each cluster face each other, and then one is rotated 60 degrees, allowing the triangles to come together as a six-pointed star, and suddenly the fourteen balls become a cube! This is the minimum stable cube formed out of spheres (Fig. 8-19). Eight spheres alone,

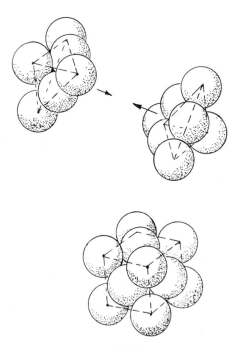

Fig. 8-19

positioned as the eight corners of the cube, are not closepacked, and that configuration would therefore be unstable, as the spheres have a tendency to roll into the unoccupied valleys.

In fact, just as a floppy toothpick cube needed six extra diagonal sticks (the six edges of a tetrahedron) to stabilize the square faces, eight balls also require an additional six, to complete a stable cube. Thus we have fourteen balls altogether: a parallel to the fourteen topological parameters (vertices plus edges plus faces) of the tetrahedron. The cube in every stable form seems to be based on an implied tetrahedron.

A Final Philosophical Note

Fuller pointed out that sphere-packing models encourage us to conceive of area and volume in terms of discrete quanta instead of as the physically impossible continuums promoted by traditional geometry:

> Because there are no experimentally-known "continuums," we cannot concede validity to the concept of continuous "surfaces" or of continuous "solids." The dimensional characteristics we used to refer to as "areas" and "volumes," which are always the *second*- and *third*-power values of linear increments, we can now identify experimentally, arithmetically, and geometrically only as quantum units that aggregate as points, both in system-embracing areal aggregates and . . . as volume-occupant aggregates. The areal and volumetric quanta of separately islanded "points" are always accountable numerically as the second and third powers of the *frequency of modular subdivision of the system's radial or circumferential vectors*. (515.011)

He argued that sphere polyhedra, having the advantage of visibly separate subunits, illustrate the otherwise invisible truth about physical reality. An awareness of the particles inherent in all physical systems (on some level of resolution) is nurtured, because in the sphere packings it is so logical to express volume and area in terms of number of units. The examples of the VE and icosahedron models demonstrate how the terms for expressing length (frequency) and area (number of particles) are actually related by a formula ($10f^2 + 2$)—as would be necessary in a new geometry.

Fuller attributed the precedent for thinking about volume in terms of quanta to Amadeo Avogadro (1776–1856) and his discovery that equal volumes of all gases, under the same conditions of pressure and temperature, contain the same number of molecules. Avogadro thereby identified volume with number of molecules a long time ago.

We take this a step further, by remembering that, although we tend to conceive of volume as a spatial continuum, our convention for quantifying an amount of space uses *number* of imaginary cubes—even if that quantity usually involves an extraneous partial cube (or fraction) tacked on to a whole number. We shall discuss the subject of volume more fully in Chapter 10. For now, we simply lay the groundwork with the evidence that polyhedra can be constructed out of a multitude of spheres, at different frequencies, and that the resulting models play an important role in satisfying Fuller's criteria for a geometry consistent with Universe.

9

Isotropic Vector Matrix

The isotropic vector matrix has already been introduced; we just didn't know its name.

If you can visualize the space-filling array of spheres in "cubic packing" described in the previous chapter, that's half the picture. Now, imagine interconnecting the centers of all spheres—and then eliminating the spheres. Two collinear radii meeting at the tangency point between adjacent spheres form one unit vector—the length of which is equal to the sphere's diameter (Fig. 9-1). The resulting array of vectors is the "isotropic vector matrix," a space-filling network of continuously alternating octahedra and tetrahedra. Reviewing the characteristics of cubic packing, we shall not be surprised to find that all the newly formed vertices (the spheres' centers) are identically situated. Two types of cells, one type of vertex.

It's not hard to see how Fuller's search for a geometry of vectors led him to the isotropic vector matrix. "Since vectors... produce conceptual structural models of energy events, and since my hypothetical generalization of Avogadro's law requires that 'all the conditions of energy be everywhere the same,'" ponders Fuller, "what does this condition look like as structured in vectorial geometry?" His answer is ready: "Obviously all the vectors must be the same length and all of them must interact [*sic*] at the same angles" (986.131b).

The isotropic vector matrix, or IVM, takes the VE a step further, consisting of identical lengths and angles, not for vectors surrounding just one point, but surrounding every point in an indefinite expanse. In Fuller's words, the IVM is "a multidimensional matrix in which the vertexes are everywhere the same and equidistant from one another" (222.25).

It is not correct to conclude that the IVM consists of many vector equilibria packed together, for the VE by itself cannot fill space. To understand why not, we look at isolated sections of the IVM. As difficult as it is to visualize the overall matrix, a single row of

Fig. 9-1. Unit vectors interconnect centers of adjacent unit-diameter closepacked spheres.

alternating tetrahedra and octahedra, or even a planar expanse, can be easily envisioned (Fig. 9-2a, b). Separate planar layers are then stacked together in such a way that every octahedron is adjacent to a tetrahedron and vice versa. Figure 9-3 shows three layers of the resulting matrix.

Every node in the IVM—as the origin of twelve unit vectors radiating outwardly—is the center of a local vector equilibrium. The ends of these unit vectors define the twelve vertices of the VE.

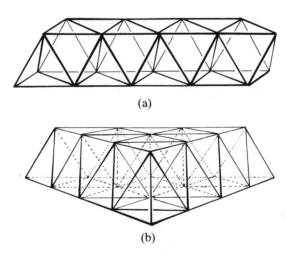

(a)

(b)

Fig. 9-2. (a) Single row of alternating tetrahedra and octahedra. (b) Planar expanse of alternating tetrahedra and octahedra.

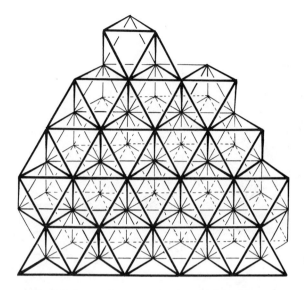

Fig. 9-3. Isotropic vector matrix and Octet Truss.

However, this does not mean that adjacent cuboctahedra pack together to produce a space-filling expanse. A symmetrical array can be created by bringing the square faces of adjacent vector equilibria together, but they are necessarily separated by octahedral cavities—framed by the triangular faces of eight converging VEs. The unavoidable octahedra between adjacent VEs provide yet another manifestation of the specificity of the shape of space. This array can be readily understood by observing in Figure 9-4 that a packing of vector equilibria is equivalent to a framework of cubes in which the corners have been chopped off, thus automatically carving out an octahedral cavity at every junction of eight boxes.

The above observations provide information about the shapes and angles of the IVM—the most symmetrical arrangement of points in space—and therefore about the shape of space itself. These characteristics reveal the basis for the term "isotropic vector matrix": in Fuller's words,

"isotropic" meaning "everywhere the same," "isotropic vector" meaning "everywhere the same energy conditions."... This state of omnisameness of vectors...prescribes an everywhere state of equilibrium." (420.01–3)

He calls the IVM "multidimensional" because it "accommodates" (or occupies) all spatial dimensions, and—consistent with his unorthodox interpretation of dimension—space is "multi-" rather than

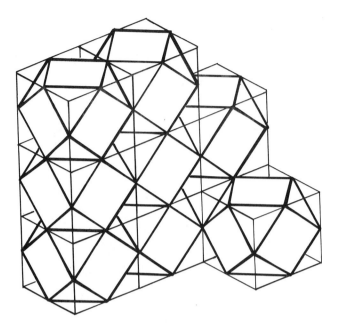

Fig. 9-4. Octahedral cavities between adjacent VEs.

"three-dimensional." Vectors are directed in every possible direction, while deliberately maintaining equivalent lengths and angles. This equivalence is necessarily determined by the symmetry of space:

> This matrix constitutes an array of equilateral triangles that corresponds with the comprehensive coordination of nature's most economical, most comfortable, structural interrelationships employing 60-degree association and disassociation. (420.01)

As seen in the earlier development of vector equilibrium, spatial "omnisymmetry" incorporates four planes of symmetry: four unique directions of equilateral triangles. Recalling the way cookies fit most economically on a baking sheet, we can feel quite comfortable with the triangular symmetry of the plane. The implication is that the shape of space can be described through four such continuous planes.

A Quick Comparison: "Synergetics Accounting"

Imagine one vertex within the IVM framework, which will be called O—for origin. A unit vector ($L = 1$) pointing in any of the twelve directions away from O ends at a vertex which we shall call A. A

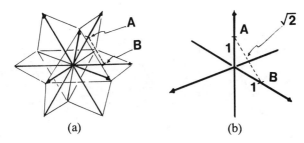

Fig. 9-5. (a) 60-degree axes. (b) 90-degree axes.

second unit vector emanating from O is given a different orientation, in a direction 60 degrees away from vector OA, and arrives at vertex B. The distance between A and B is also unit length (Fig. 9-5a). As simple and repetitive as this observation might seem, it is the essence of "synergetics accounting," as opposed to "algebraic accounting."

The same procedure applied to a 90-degree framework, also using unit vectors, places vertices A and B an irrational-number distance apart. Unit increments along the x and y axes create points labeled simply $\{(0, 1)$ and $(1, 0)\}$, which are separated by a (not so simple) irrational $\sqrt{2}$ or $1.41421\ldots$ units. Furthermore, if pathways between vertices are to follow along the network of vectors, the square grid disallows the shortest route between A and B. Observe in Figure 9-5b that to get from A to B along the prescribed grid requires traveling two units, despite the fact that they are separated by only $1.414\ldots$ units. As energy always takes the shortest route, argues Fuller, the XYZ system clearly does not serve to illuminate the events of physical reality. In contrast, the most expedient route from A to B in the triangular grid happens to be directly along the unit-length vector connecting the two points.

Irresolvable numbers do exist within the IVM, as there are square cross-sections (corresponding to the square faces of the VE); however, the irrationals are not part of the fundamental *orientation* of the system. To Fuller, a simple procedure, like the one described above, ought to yield simple ("omnirational") results. If a frame of reference is itself convoluted, its ability to describe and measure other phenomena will be all the more so.

Cells: "Inherent Complementarity"

Let's back up and start again. The goal is to establish a symmetrical and complete spatial array of vectors, and one logical approach might be to start with space's minimum system. We gather a number

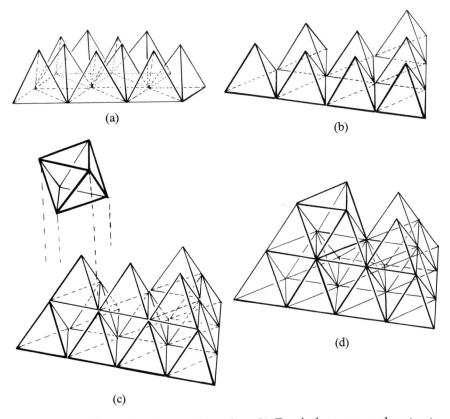

Fig. 9-6. (a) Tetrahedra do not fit together. (b) Tetrahedra rearranged vertex to vertex. (c) A second row of tetrahedra is placed above the first, such that all tetrahedra meet vertex to vertex; the arrangement automatically creates octahedral cavities. (d) Alternating tetrahedra and octahedra can fill space indefinitely.

of unit-vector tetrahedra and place them on the ground side by side, and it is immediately apparent that they will not pack together to produce a continuous expanse (Fig. 9-6a). Awkward gaps between adjacent tetrahedra cannot be filled by regular, or symmetrical, shapes, precluding an isotropic array. This constraint is not new: we saw in the previous chapter that tetrahedra cannot fill space; however, vector models display the shapes more clearly than sphere packings.

Still in pursuit of a space-filling array, we now rearrange the tetrahedra so that they meet vertex to vertex, each with one edge along a continuous line, in multiple adjacent rows (Fig. 9-6b). Notice what happens when we interconnect the tetrahedral peaks (as would be the result if a second layer of tetrahedra were placed on top of the first): precise octahedral cavities emerge in between all tetrahedra, automatically completing the isotropic vector matrix with its alternating two shapes (Fig. 9-6c, d).

The reverse is also true: octahedra cannot themselves fill space, but when arranged edge to edge (not face to face or vertex to vertex) the emergence of by-product tetrahedra reconfirms the persistent pairing. The developed matrix with its unit vectors and equivalent points of convergence thus depicts the *inherent complementarity* of space—meaning inevitable co-occurrence of octahedra and tetrahedra. Fuller draws a parallel between this and other inseparable pairs such as electron–proton, concave–convex, male–female, and tension–compression. "Inherent complementarity of Universe" applies to the entire phenomenon of interdependent partners, whether atomic constituents or polyhedral space-fillers. (Chapter 12 will further elaborate on Fuller's interpretation of the significance of "inherent complementarity.")

In conclusion, the development of both the VE and the IVM—whether through closest packing of spheres or by symmetrical arrangement of vectors—supports a sense of the balance of octahedral and tetrahedral symmetries inherent in space. Both configurations build *themselves*—in response to spatial constraints.

A Complete Picture

We now step inside the IVM to complete our investigation of this omnisymmetrical network of vectors. The centers of closepacked spheres constitute the vertices of most regular and semiregular polyhedra. We looked at some of them in the previous chapter, and with the use of toothpicks instead of Ping Pong balls, the outlines of these shapes can be more easily discerned.

We have already observed that vertices in the IVM fall into triangular patterns in four distinct planar directions. Through our experience with cubic packing, we know to look for an additional three planes of symmetry, characterized by a square distribution of vertices. Neighboring octahedra share the edge between them, and thus the cross-sections of individual octahedra join together, forming the square pattern of graph paper in three orthogonal directions. Figure 9-7 highlights an IVM squared plane, by omitting certain lines; half octahedra shown without tetrahedral edges clarify the square aspect of the omni-triangulated matrix.

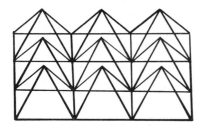

Fig. 9-7. Half octahedra shown without tetrahedral edges, to emphasize a square cross-section of the IVM.

Angles

The combination of these two simple shapes in the IVM yields surprisingly many different angles and potential shapes. Our attention tends to be focused on the surface characteristics of the tetrahedron and octahedron, and so we observe only triangles and 60-degree angles. However, the interior structure introduces distinct new elements, such as the square octahedral cross-sections discussed above. The next step is to list other interior angles, for a sense of the range of possible shapes contained within the matrix.

The dihedral angle (angle between two faces) in a regular tetrahedron is approximately 70°32'. The tetrahedron is unique in that any two edges at a given vertex are part of a common face. Every other polyhedron has interior angles in addition to surface angles between edges, thus adding to the range of shapes incorporated into each system. For example, any two nonadjacent edges at an octahedron vertex meet at 90-degree angles, thus forming square cross-sections. The octahedron dihedral angle is 109°28', which—as the supplement[1] to the tetrahedron's 70°32'—results in perfectly flush surfaces shared by adjacent octahedra and tetrahedra, allowing the continuous planes of the IVM (Fig. 9-8).

Both dihedral angles at first appear to be such irregular numbers that this exact geometric fit is surprising—especially when we recall our first encounter with the two shapes. Remember that we simply surrounded vertices by three triangles, then by four, allowing the systems to close off with as many triangles as necessary. (Refer to Chapter 4.) The process provided no basis for predicting the exact

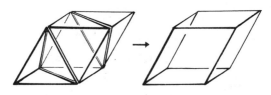

Fig. 9-8

complementarity of the two polyhedral systems. The coincidence continues with the addition of central nodes, as we shall see below.

Locating New Polyhedral Systems

The first new polyhedron consists simply of one octahedron with a tetrahedron on two opposite sides (Fig. 9-8). The result, a *rhombohedron*, can be seen as a partially flattened cube. A toothpick–marshmallow model demonstrates the transition effectively, because the marshmallow joints have sufficient stiffness to hold either inherently unstable shape. The rhombohedron's direct relationship to the cube suggests a space-filling capability, which we shall explore in greater depth in Chapter 12.

The next candidate, the VE, is too familiar to warrant further description at this point. Twelve cuboctahedral vertices can be

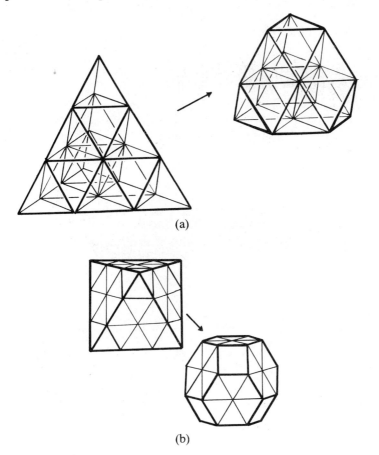

(a)

(b)

Fig. 9-9. (a) Truncation of three-frequency (3v) regular tetrahedron. (b) Truncation of 3v regular octahedron, showing only external surface of system.

located around every point in the IVM, thereby embracing eight tetrahedra and six half octahedra.

Furthermore, higher-frequency versions of any of the above polyhedra—tetrahedron, octahedron, rhombohedron, and VE—can be easily located within the matrix, thus establishing the foundation for truncated polyhedra. Subtract a half octahedron from each of the six corners of a three-frequency octahedron to yield a symmetrical "truncated octahedron" with fourteen faces: six squares and eight regular hexagons (Fig. 9-9b). A "truncated tetrahedron," with four hexagons and four triangles, is left after a single-frequency tetrahedron is removed from each corner of a three-frequency tetrahedron (Fig. 9-9a). The same procedure applies to higher-frequency versions of any of the above shapes, as well as further truncations of truncated shapes. Such transformations can be plotted indefinitely.

Duality and the IVM

We now introduce a new level of flexibility with the addition of a new set of vertices—in the exact center of each tetrahedron and octahedron. These new vertices are connected to the original IVM vertices, thereby introducing radial vectors into each of the original cells. Figure 9-10 shows a single octahedron and tetrahedron with these central nodes.

Angles

The central angles of a tetrahedron is approximately 109°28′, exactly equal to the octahedron's dihedral angle. No longer surprised by such relationships, we go on to look inside the octahedron and note

Fig. 9-10. Central nodes of tetrahedron and octahedron.

its central angle of 90 degrees, which is the surface angle of a cube. The octahedron's three body diagonals, or six radii, thus form the *XYZ* axes. Figure 9-10 highlights these central angles by showing these two polyhedra with central nodes and radii. Right angles are thus integrated into the IVM system as by-products of the (stable) triangulated octahedron, rather than by an arbitrary initial choice of a network of unstable cubes. Since the IVM complex of octahedra and tetrahedra emerges automatically as a consequence of its unique property of spatial omnisymmetry, the array is not the product of an arbitrary choice.

Polyhedra

We now observe considerable expansion of our inventory of generated shapes. Starting with the most familiar, we isolate the minimum cube. Formed by a single tetrahedron embraced by four neighboring eighth-octahedral pyramids, or *octants*, the cube is once again based on the tetrahedron. We first encountered this relationship in "Structure and Pattern Integrity" (using the tetrahedron to establish the minimum stable cube), and now we have determined the exact shape of the leftover space: four eighth-octahedra. This observation indicates that "degenerate stellation" of the tetrahedron forms a cube. The four vertices of the tetrahedron, together with the four centers of neighboring octahedra, provide the eight corners of this basic building block. Its six square faces are created by two adjacent quarters of the square cross-sections of single-frequency octahedra (Fig. 9-11). As with other IVM systems, larger and larger cubes will be outlined by more remote octahedron centers.

Next, we embrace a single octahedron by eight quarter tetrahedra, thereby outlining the rhombic dodecahedron, whose twelve diamond faces have obtuse angles of 109°28′ and acute angles of 70°32′—generated by the tetrahedral central angle and two adjacent axial angles, respectively. Its eight three-valent vertices are the centers of embracing tetrahedra, while its six four-valent vertices are the original octahedron vertices (Fig. 9-12). Once again, we observe the relationship of duality between the VE and rhombic dodecahedron. The former has fourteen faces (six four-sided and eight three-sided) corresponding to the four-valent vertices and three-valent vertices of the latter. Likewise, the twelve four-valent VE vertices line up with the twelve rhombic faces. (Refer to Fig. 4-14.)

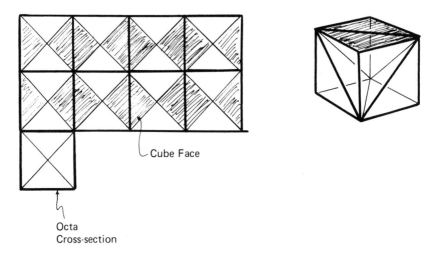

Cube Face

Octa
Cross-section

Fig. 9-11. Relationship of cube and IVM.

Domain

The duality between VE and rhombic dodecahedron illustrates the relationship of duality and domain. Having already seen that spheres in closest packing outline the vertices of the VE, we now turn our attention to the *domain* of individual spheres.[2] The domain of a sphere is defined as the region closer to a given sphere's center than to the center of any other sphere. This necessarily includes the sphere itself, as well as the portion of its surrounding gap that is closer to that sphere than to any other. Imagine a point at the exact center of an interstitial gap; this will be the dividing point between neighboring domains, that is, a vertex of the polyhedron outlined by the sphere's domain. This domain polyhedron happens to be the rhombic dodecahedron. As each sphere in cubic packing is by definition identically situated, each domain must be the same. Therefore, the shape of this region consistently fits together to fill space. Fuller's term for the rhombic dodecahedron is "spheric" because of this relationship to spheres in closest packing.

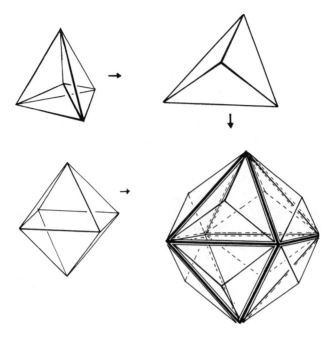

Fig. 9-12. Degenerate stellation of the octahedron is accomplished by affixing quarter tetrahedra to each face.

We now have an experiential basis for the VE–rhombic-dodeca-hedron duality. Twelve vectors emanate from any point in the IVM, locating the *vertices* of the VE, while poking through the center of the twelve diamond *faces* which frame the point's domain. We were introduced to duality as exact face-to-vertex correspondence, and now we see how duals can be instrumental in locating a system's domain. Our investigation of space-filling in Chapter 12 will explore this relationship more fully.

Returning to the IVM, we observe that four rhombic dodecahedra, or "spherics," come together at the center of each tetrahedron, such that the tetrahedron central angle becomes the obtuse surface angle of the spheric. In the same way, eight cubes meet at the center of each octahedron, as allowed by the shared 90-degree surface and central angles, respectively.

For clarity, we shall refer to the new network, interconnecting the centers of all octahedral and tetrahedral cells, as IVM′, and we can draw the following conclusion. If the vertices of a given polyhedron are located in the IVM, then that system's dual will be outlined by the IVM′, and vice versa. Similarly, if a polyhedron is centered on a vertex of the IVM, its dual will be centered on a vertex in IVM'. For example, we recall our first case of duality: the vertices of the octahedron's dual, the cube, are supplied by octahedron centers, which are nodes of IVM'.

This discovery leads us to another assumption. As truncation of our familiar polyhedra yields shapes contained within the IVM, the dual operation, stellation, should produce polyhedra outlined by IVM'. The assumption is valid: the additional IVM' vertices provide the loci for the vertices of stellated versions of these basic shapes. Actually, this observation is not new, for we have already seen that quarter-tetrahedral pyramids affixed to octahedron faces produce Fuller's spheric, or, in other words, that a degenerately stellated octahedron becomes a rhombic dodecahedron. The three-valent vertices of this diamond faceted shape are tetrahedron centers, by definition nodes of IVM'.

Framework of Possibility

The isotropic vector matrix gives us a description of the symmetry of space. We can think of this matrix as a framework of possible directions and configurations of ordered space, or more simply, as a *frame of reference*. It is a network of vectors specifically situated to model nature's eternal tendency toward equilibrium. Lines are forces, length is magnitude, and all is in balance. The IVM weaves together a number of synergetics ideas: minimum system of Universe, vector equilibrium (both exhibiting four planes of symmetry), twelve degrees of freedom, complementarity of octahedra and tetrahedra, space-filling, and stability (exclusively a product of triangulation). In so doing, it sets the stage for an energetic mathematics, and systematizes further investigation.

The IVM also provides an alternative to the *XYZ* system's absolute origin. Every vertex in the IVM can be considered a temporary local origin, which, as reinforced by Fuller's use of the concept of "systems," is consistent with the requirements of describing Scenario Universe. ["All points in Universe are inherently centers of a local and unique isotropic-vector-matrix domain..." (537.11).] There can be no "absolute origin" in a scenario.

Finally, by describing such a wide variety of ordered polyhedra—and thereby clarifying the relationships between different shapes—the IVM supports Fuller's concept of "intertransformability." Countless potential shapes and transformations can be systematically represented within this omnisymmetrical matrix; it is a framework of possibility.

Invention: Octet Truss

Our familiarity with the IVM enables us to visualize and appreciate Fuller's "Octet Truss." Awarded U.S. Patent 2,986,241 in 1961, this structural framework is so widespread in modern architecture that one might assume buildings have always been constructed that way. Again, as the story goes, the invention can be traced to 1899 when Bucky was given toothpicks and half-dried peas in kindergarten. So extremely farsighted and cross-eyed that he was effectively blind (until he received his first pair of eyeglasses a year later), Bucky Fuller did not share the visual experience of his classmates and therefore lacked the preformed assumption that structures were supposed to be cubical. Thus, as other children quickly constructed little cubes, young Bucky groped with the materials until he was satisfied that his structures were sturdy. The result, much to the surprise of his teachers (one of whom lived a long, long life, and periodically wrote to Fuller recalling the event) was a complex of alternating octahedra and tetrahedra. He had built his first Octet Truss—also the first example of what was to become a lifetime habit of approaching structural tasks in revolutionary ways.

The experience had a great impact on the four-year-old, as he recounted in a 1975 lecture:

> All the other kids, the minute they were told to make structures, immediately tried to imitate houses. I couldn't see, so I felt. And a triangle felt great! I kept going 'til it felt right, groping my way.... [3]

The truss's omnisymmetrical triangulation distributes applied forces so efficiently that the resulting strength of such an architectural framework is far greater than predicted by conventional formulae:

> The unitary, systematic, nonredundant, octet-truss complex provides a total floor system with higher structural performance abilities than engineers could possibly ascribe to it through conventional structural analysis predicated only upon the behavior of its several parts. (650.11)

Struts can be all one length, thus simplifying construction, while the minimal volume-to-material ratio inherent in the geometry of the tetrahedron[4] maximizes resistance to external loads. The intrinsic stability of triangulation together with efficient dispersal makes this system the most advantageous possible use of materials in a space-frame configuration:

> It is axiomatic to conventional engineering that if parts are "horizontal," they are beams; and the total floor ability by such conventional engineering could be no stronger than the single strongest beam in the plural group. Thus their prediction falls short of the true behavior of the octet truss by many magnitudes.... (650.11)

The octet truss takes the conceptual matrix into physical realization, and thus embodies Fuller's design science concept of using geometric principles to human advantage.

We can now appreciate the difference between diamond and graphite. Both consisting of carbon atoms, the former is exquisitely hard and clear, the latter soft and grey, and their differences are due to geometry. Carbon atoms in the structure of diamond take advantage of the strength of tetrahedra; their organization can be thought of as a double octet truss, two intersecting matrices with the vertices of one overlapping the cells of the other. Stabilized by the high number of bonds between neighboring atoms, which also allow forces to be distributed in many directions at once, the configuration is supremely invulnerable. In contrast, carbon atoms in graphite are organized into planar layers of hexagons—triangulated and stable in themselves, but not rigidly connected to other layers. As a result, separate layers are able to shift slightly with respect to each other, which does not mean that graphite lacks all stability, but rather that it is relatively soft. This softness enables graphite to leave visible residue on the surface of paper, thus performing its useful function in pencils. A more illustrative although less widely recognized application is that these sliding layers make graphite a powerful lubricant. The comparison provides a spectacular example of synergy: rearrangement of identical constituents produces two vastly different systems.

Thus we see that nature also employs design science.

10

Multiplication by Division:
In Search of Cosmic Hierarchy

It may seem that we have strayed from Fuller's "operational mathematics" while investigating the symmetrical properties of various polyhedra in the previous chapter. Recall that "operational" indicates an emphasis on procedure and experience: what to *do* to develop and transform models or systems. "Multiplication by division" brings us back to experience, introducing an operational strategy, which will add new meaning to Fuller's term "intertransformabilities." We thus elaborate on the shared symmetries among shapes while discovering new transformations from one to another, and this time previous experience allows us to anticipate results.

Multiplication by division describes Bucky's journey through our expanding polyhedral inventory. Previous exposure to both Loeb's work and the IVM sets the stage, making us so familiar with these shapes that additional results can be immediately placed in context. The transformations explored in this chapter occur within the IVM frame of reference, adding volume relationships to our accumulated information about topology and symmetry. You may be surprised to find that many statements seem obvious at this point; resist the temptation to dismiss them as trivial. Appreciate instead the implication—which is that we cannot take a wrong turn. Each step is inherently tied to the shape of space; we can only uncover what is already there.

Volume

The use of ratio is an inherent part of quantifying volume, and yet not everyone is aware of the implicit comparison. As with measuring distance, our conventional units can seem like a priori aspects of volume.

Once again, Fuller calls our attention to Avagadro's discovery that a given volume of any gas, subject to identical conditions of temperature and pressure, always contains the same number of molecules. "Suddenly we have volume clearly identified with number," declares Bucky.

Actually, volume is intrinsically related to number. When we ask, "what is the volume of that swimming pool"? we expect an answer expressed in terms of some number of "cubic feet." What this answer tells us is how many cubes with an edge-length of one foot could fit into the pool. Whether the situation calls for feet, inches, or centimeters, a cube of unit edge length is conventionally employed as one unit. The word "volume" may evoke an image of a continuum; however, it is quantified in terms of discrete quanta. We so uniformly express spatial quantity in terms of cubes that we are simply not aware of the invisible framework of "ghost cubes" incorporated into our concept of volume. We conceptualize space cubically: length, width, and depth seem absolutely fundamental directions. Again Bucky points out that this conceptual cube is a remnant of flat-earth thinking. Myopic in cosmic terms, humanity readily adopted the orthogonal box as the correct shape with which to segment space.

Results: Volume Ratios

Volume has to be measured relative to *something*, so why not experiment with the tetrahedron? We are so used to using the cube, the suggestion seems blasphemous—a violation of basic laws of volume. Nevertheless, given our growing list of the tetrahedron's unique properties, such an experiment might be worthwhile.

Accordingly, we allow a tetrahedron of unit-edge length to be called one unit of volume. The results are astonishingly rewarding. Perfect whole-number values describe the volumes of most of the polyhedra covered so far, and *all* of those contained within the IVM and IVM′ combined. (Some exceptions are found in transition shapes—those that fall in between IVM vertices—as we shall see in the next chapter.) In contrast, the volumes of these familiar polyhedra, relative to a *cube* as the unit shape, are strangely cumbersome values—often irrational (never-ending) decimal fractions. Table V displays the results, which we shall derive below. It compares the volume ratios generated by three different polyhedra successively adopted as one unit of volume. Five different systems are compared first with the unit-edge cube, then with the unit-diagonal cube, and finally with the unit-edge tetrahedron.

Table V. Volume Ratios

| Polyhedron Measured | Polyhedron Taken as Unit of Volume: | | |
| | Cube | | Tetrahedron |
	Unit Edge	Unit Diagonal	
Tetrahedron	0.11785	0.33333	1
Octahedron	0.47140	1.33333	4
Cube			
(unit diagonal)	0.35356	1	3
Rhombic			
dodecahedron	0.70710	2	6
VE	2.35700	6.66666	20

Remember that these polyhedra arose as a consequence of spatial symmetry; we simply located vertices within the unique isometric array of vectors. Recalling this origin, it is again clear that the various shapes and sizes of the polyhedra in question are not the product of deliberate design. Their whole-number volume ratios are not contrived; we stumble onto them after the fact.

Why investigate two different cubes? Primarily, to demonstrate that neither choice yields the elegant results disclosed by the tetrahedron. All factors considered, the unit-diagonal cube is a better choice, for it arises naturally out of the IVM network—that is, out of the shape of space! Further justification for this choice will be developed below.

Imagine building a cube out of the requisite twelve struts. The topological recipe, which simply calls for three-valent vertices and four-valent faces, in no way indicates precisely what the finished product should look like. Without deliberate shaping into a perpendicular form by a knowing hand, the configuration does not favor any particular surface angles. Which version of this hexahedron of quadrilaterals is the desired result? This ambiguity must somehow be resolved. Chapter 5 revealed that an orthogonal "cube"—as defined by mathematics—cannot be reliably created without diagonal braces. (Refer back to Fig. 5-3.) Compare this experience with building an octahedron out of the same twelve structs. Following the recipe of four-valent vertices and three-valent faces. the octahedron builds itself. The interior shape is precisely specified by its topology; in other words, the procedure leaves no room for choice. In order for the cube to have that kind of integrity, or exactitude, six face diagonals must be inserted. An inscribed tetrahedron solves the problem in a single step.

The above comparison reinforces our previous experience of how the right-angled cube fits into spatial symmetry. Fuller's "operational mathematics" prescribes learning by procedure: pick up a box of toothpicks (pre-cut unit-vector models) and start building. Space will let you know what works. The satisfaction gained from feeling the cube hold its shape draws attention to its supporting diagonal members, and thus unit-vector diagonals seem an appropriate choice for comparing volumes. As volume is always a matter of ratio, we want to insure that our comparisons make sense, in this case, remain consistent with space's isotropic vector matrix. That the vector-diagonal cube is contained within this hierarchy adds to the advantages established by its stability.

Shape Comparisons: Qualities of Space

Once again, we take advantage of the ease of working with planar configurations, before tackling space. We thus start by comparing the characteristics of triangles and quadrilaterals, and then we shall attempt to apply our conclusions to tetrahedra and cubes. We begin by drawing an irregular version of each polygon, and observe the following. If we bisect the edges of the two figures and interconnect these points as shown in Figure 10-1a, both shapes are divided into four regions. However, the triangle and quadrilateral exhibit a strikingly different result. A triangle, no matter how irregular, automatically subdivides into four identical triangles—all *geometrically similar* to the original, that is, the same shape but a different size. Observe in Figure 10-1a that this is not true for the quadrilateral. Excluding the special case of a parallelogram, the four small quadrilaterals will not be similar to their framing shape.

Now, on to space! A tetrahedron (of any shape or size) carved out of firm cheese can be sliced parallel to one of its faces, removing a slab of any thickness, to produce a new smaller tetrahedron with precisely the same shape as the original (Fig. 10-1b). This does not work with the cube, or for that matter, with any other polyhedron, regular or not. The ability to "accommodate asymmetrical aberration" without altering shape, observes Fuller, is unique to the minimum system of Universe. We add this observation to a growing list of special properties of the tetrahedron. (Appendix D.) Similarly, as will be demonstrated below, an irregular tetrahedron can be subdivided to create smaller identical tetrahedral shapes, whereas an irregular hexahedron will yield dissimilar hexahedra, in the same manner as its planar counterpart, the quadrilateral. Evidence thus gradually accumulates to support using the tetrahedron (instead of

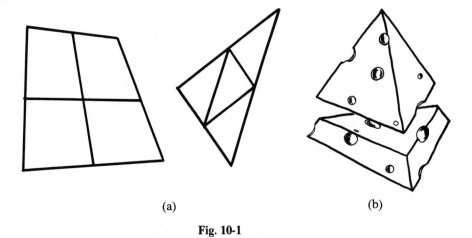

(a)   (b)

Fig. 10-1

the cube) as the basic unit of structure—or mathematical starting point.

Volume: Direct Comparison

Before looking more closely at volume ratios, we review the following mathematical generalization. No matter what unit of measurement is employed, the volume of any container is mathematically proportional to a typical linear dimension raised to the third power. This means that if we have two geometrically similar polyhedra, one with twice the edge length of the other, the larger will contain exactly eight times the volume of the smaller. (Having the same shape, the two systems will share a common "constant.")

To bring this mathematical law into experiential grasp, we consider two familiar shapes. It is easy to visualize that a cube of edge length 2 consists of eight unit cubes (Fig. 10-2). Now imagine a tetrahedron of edge length 2. Observe in Figure 10-3 that, just like its cubic counterpart, the altitude of a two-frequency tetrahedron is two

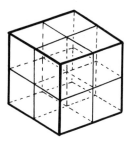

Fig. 10-2. Two-frequency cube consists of eight unit cubes.

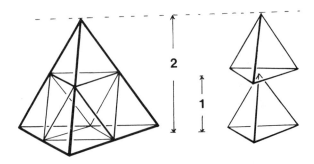

Fig. 10-3

times that of a unit tetrahedron, and similarly that each face subdivides into four unit triangles. The latter observation indicates that the area of the large tetrahedron's base is four times that of the small tetrahedron's unit-triangle base. Emulating the approach employed by Loeb in his "Contribution to Synergetics,"[1] we can deduce the following, simply by utilizing traditional geometric formulae.

Let Vol_T and Vol_t represent the volumes of the large (two-frequency) and small (unit-length) tetrahedra; A_T and A_t, the areas of their bases: and H_T and H_t, their altitudes.

According to the formula

$$\text{volume of pyramid} = \text{constant} \times (\text{area of base}) \times \text{height}.$$

So

$$\text{Vol}_T = KA_TH_T \quad \text{and} \quad \text{Vol}_t = KA_tH_t,$$

and since we observed in Figure 10-2b that the base of the large tetrahedron is divided into four triangles, each of which is equal to the base of the small one, it is clear that $A_T = 4A_t$. Similarly, the altitude of the larger pyramid is twice that of the smaller, or $H_T = 2H_t$. Substituting, we have

$$\text{Vol}_T = K \times 4A_t \times 2H_t,$$

or

$$\text{Vol}_T = 8KA_tH_t,$$

and since

$$\text{Vol}_t = KA_tH_t,$$

it follows that

$$\text{Vol}_T = 8\,\text{Vol}_t,$$

or, in words, that the volume of the big tetrahedron is eight times that of the little tetrahedron. The constant K cancels out of the expression when the two equations are compared. This conclusion will be useful in deriving the volume ratios displayed in Table V.

Multiplication by Division

Multiplication occurs only through progressive fractionation of the original complex unity of the minimum structural systems of Universe: the tetrahedron. (100.102b)

Instead of starting with parts—points, straight lines, and planes—and then attempting to develop these inadequately definable parts into omnidirectional experience identities, we start with the whole system in which the initial "point"... inherently embraced all of its parameters... all the rules of operational procedure are always totally observed." (488.00)

Deeply impressed by Arthur Eddington's definition of science as "the systematic attempt to set in order the facts of experience," Fuller constantly sought meaningful organizations for groups of experiences or events. "Multiplication by division" is one such effort. ("Events" of course includes structures and almost anything else; in energetic Scenario Universe, things are events.)

Essentially, "multiplication by division" derives volume relations through the straightforward logic of direct observation, rather than by rote application of traditional formulae—which lead us through awkward values before revealing the underlying simple relationships. As will be seen below, this direct observation is accomplished by comparing given polyhedra to the unit-length tetrahedron.

Tetrahedron as Starting Point

Fuller's organizing strategy begins with the tetrahedron, because as the "topologically simplest structural system," it is a logical starting point. Consistent with his emphasis on "whole systems," the ulti-mate reference point in synergetics is "Universe." The tetrahedron thus acts as an appropriate "whole system" for the procedure described below, in that it is "the first finite unitarily conceptual subdivision of... Universe" (987.011b). More complicated systems are developed through subdivision of this tetrahedral starting point, so that a progression is contained within (and organized by) the whole:

In respect to such a scenario Universe multiplication is always accomplished only by progressively complex, but always rational, subdivisioning of the initially sim-plest structural system of Universe: the sizeless, timeless, generalized tetrahedron. (986.048b)

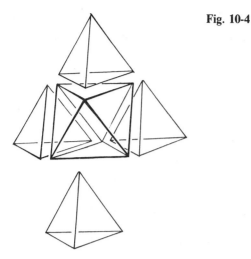

Fig. 10-4

Onward! We seek to develop a variety of polyhedra through subdivision of the "whole" and in so doing provide the "experimental evidence" to verify the results shown in Table V. We imagine a single regular tetrahedron, and then bisect each edge to create the two-frequency tetrahedron shown in Figure 10-3. One by one, we remove a single-frequency tetrahedron from each of the four corners, unwrapping the hidden octahedron (Fig. 10-4). Choping off four unit tetrahedra subtracts four units of volume from the initial total of eight (the value determined earlier for a double-edge-length tetrahedron), indicating that the octahedral remainder has a volume of exactly four. In other words, an octahedron has four times the volume of a tetrahedron of the same edge length. We thus begin to derive the values in Table V.

Cube

Next, we split the octahedron in half, separating the two square-based pyramids. Two additional perpendicular slices divide each pyramid into quarters (Fig. 10-5), producing eight sections, or "octants," each with a volume of one-half (a volume of four, divided by eight, is equal to one-half). Each octant is an irregular tetrahedron with a unit-length equilateral base and three right-isoceles-triangle sides. The perpendicular corners of the eight octants meet at the octahedral center of gravity, an orthogonal relationship first noted in the previous chapter when IVM' vertices (body center) were added to the cells of the IVM.

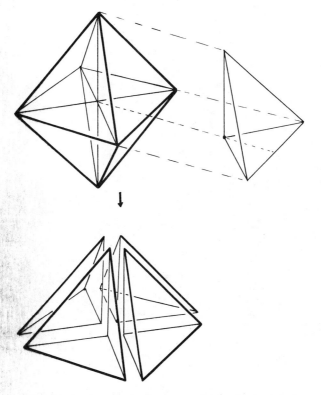

Fig. 10-5. Derivation of an octant: one-eighth of an octahedron.

To reconfirm the octahedron–tetrahedron volume relationship, we place an octant (with its equilateral face down) next to a regular tetrahedron on a flat surface and observe that the altitude of the octant is exactly half that of the tetrahedron. (A second octant can be put on top of the first to check: the height of both octants together is equal to the altitude of the tetrahedron, as shown in Figure 10-6.) An octant, therefore, has the same base and half the altitude as a regular tetrahedron, reconfirming that its volume is exactly half the volume of the tetrahedron.

In "Structure and Pattern Integrity" we discovered that a regular tetrahedron fits inside a cube, and subsequently we learned that each of the four "leftover" regions is equivalent to the portion of an octahedron from one face to its center of gravity (Chapter 9). We now take advantage of the recently disassembled octahedron for an experiment. Having equilateral triangles in common, octants can be superimposed on each face of a unit tetrahedron. One by one, four octants surround and thus obscure the tetrahedron. Lo and behold, a

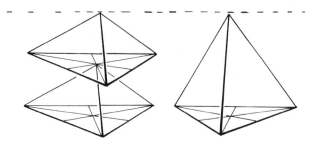

Fig. 10-6. The altitude of an octant is equal to one-half the altitude of the tetrahedron.

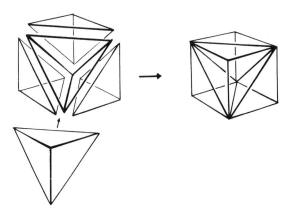

Fig. 10-7. Four octants added to one tetrahedron produce one cube.

perfect cube emerges (Fig. 10-7). Four octants, with a combined volume of two units, have been added to the unit-volume tetrahedron, for a total volume of three. Compared again with the irregular volumes listed in Table V, generated by the unit-edge cube, these whole-number ratios for the cube and octahedron seem especially remarkable.

Vector Equilibrium

The volume of the VE can be quickly determined through direct observation. Recall that twelve unit-length radii outline eight regular tetrahedra and six half octahedra. This fact, combined with our newly generated volume data, provides a conclusive value for the vector equilibrium: six half octahedra, each with the tetrahedron volume of two, plus eight unit-tetrahedra yields a total tetrahedron volume of twenty: 12 + 8 = 20. This simple breakdown supplies further evidence of a natural order of precise volume relationships, and it is especially reassuring that a unit-length vector equilibrium—the conceptual foundation of Fuller's energetic mathematics—falls into place with its own whole-number volume ratio.

Rhombic Dodecahedron

To begin with, our recently disassembled octahedron must be put back together. Octants are thus lifted away from the composite cube shown in Figure 10-6, and their right-angled corners are again turned inward to meet at the octahedron's center. Next, we divide two tetrahedra into quarters. Each tetrahedron yields four shallow pyramids, encompassing the region from an outside face to the center of gravity (Fig. 10-8). As before, the equilateral base of each shallow pyramid allows the quarter tetrahedra to fit directly onto an octahedral face, and as soon as eight quarter tetrahedra are attached to the octahedron, a rhombic dodecahedron emerges (Fig. 9-12). Two units of volume have been added to the octahedral four, for a total of exactly six.

Looking at the shape of the rhombic dodecahedron, with its strange angles and facets, the perfection of this whole-number volume

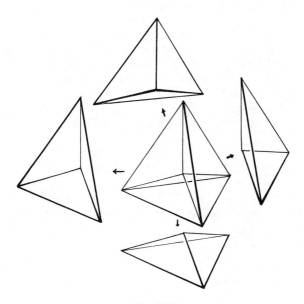

Fig. 10-8

relationship seems particularly remarkable. Because both shape and topological characteristics of the tetrahedron and rhombic dodecahedron appear utterly dissimilar, the discovery of this orderly relationship between the two polyhedra adds significantly to our growing sense of an underlying spatial order.

Again, the considerable flexibility of the IVM framework enables us to plot all of the above polyhedra with IVM and IVM' vertices. The rationale for using the unit-diagonal cube also applies to the rhombic dodecahedron, which arises naturally out of the interaction of IVM and IVM' cells. Recall that every octahedron in the matrix is surrounded by quarter tetrahedra, thereby defining rhombic dodecahedra with unit-length diagonals.

We discovered many shared symmetries in the previous chapter by dissecting the IVM, and we now develop the significance of these observations with the discovery of the rational volume relationships inherent in this framework. Bucky was not the first to discover these ratios, but he may have been their most visible spokesman. He brought this esoteric information to the attention of countless packed lecture halls, as one of the more satisfying indications of the fallibility of our coordinate system. (Such sublime disclosures by nature must not go unheralded!) These volume ratios provided Fuller with a powerful source of confidence in the legitimacy of pursuing synergetics, and indeed their significance is worth our serious consideration. What accounts for the lack of attention paid to these simple mathematical facts? Loeb offers the following explanation:

> When these relations are derived with the aid of the usual formulae for the volume of a pyramid, $V = \frac{1}{3}Ah$, a good many irrational numbers are involved, and the simple integral ratios emerge almost incidentally. Somehow, these simple integral values of the volume ratios of common solids are not part of our scientific culture, and a lack of familiarity with them frequently leads to unnecessarily cumbersome computations. *It appears that a bias of our culture to orthogonal Cartesian coordinates has obscured these relations.* [My italics.][2]

Multiplication by Division

Bisecting the edges, as before, we take special note of the tetrahedron's square cross-section. This fourfold symmetry was a significant factor in previous discussions, notably in the sphere-packing demonstration, in which two sets of spheres came together at 90 degrees and unexpectedly produced a tetrahedron. However, this aspect of the tetrahedron is easily overlooked; as a triangular pyramid, with its preponderance of 60-degree angles, this shape is easily perceived as a completely triangular affair.

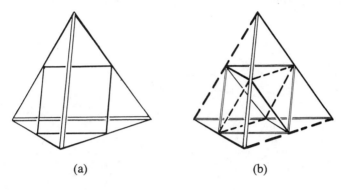

(a)   (b)

Fig. 10-9

Suppose we ask the following question: how much of our cheese tetrahedron would be chopped off when the newly exposed surface (created by the slice) is a perfect square? Without a certain amount of previous exposure to the tetrahedron, your reaction would probably be that the slicing-plane could *never* be square. It might be a very small triangle, or any number of larger triangles as you position the knife closer to the base. But a square?!

Wait. The tetrahedron has four faces. Do they somehow outline a square? Anyone who has read this far knows the answer, but countless students challenged to find that hidden square have been stuck. Handicapped by the perpendicular bias of mathematics, they are unable to find the square cross-section in the exact center of the tetrahedron, which—once seen—is unmistakable (Fig. 10-9a).

The square in Figure 10-9a is parallel to and between two opposing edges, which themselves are perpendicular to each other. Delineating the square cross-sections corresponding to each of the three sets of opposite edges, this aspect of the tetrahedron's symmetry is exhausted, and the two-frequency subdivision is complete. A total of twelve new edges outline the octahedron, and by now this relationship is quite familiar (Fig. 10-9b).

We continue inward. This time, bisect and interconnect the edges of the *octahedron*. The process is equivalent to Loeb's "degenerate truncation" and outlines the edges of a vector equilibrium hiding inside the regular octahedron (Fig. 10-10b). We could continue, by joining the midpoints of VE edges, to produce a "rhombicuboctahedron"; however, Fuller's sequence comes to an end at the vector equilibrium. The final lines, which are one-quarter the length of the edges of the original tetrahedron and the smallest vectors in the model, are thus designated as unit vectors.

We quickly run through the sequence in reverse to review the geometric relationships. A four-frequency tetrahedron, in which each

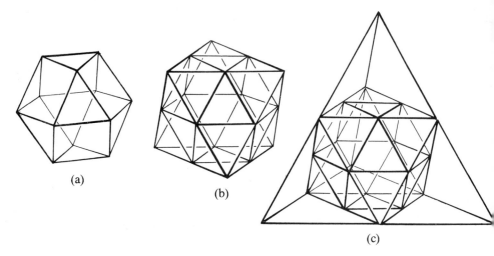

(a)

(b)

(c)

Fig. 10-10

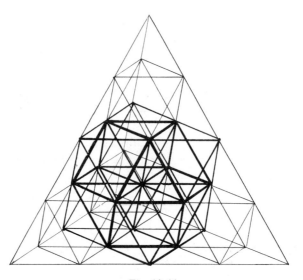

Fig. 10-11

edge is equivalent to four unit vectors, is the smallest tetrahedron to contain a complete VE in its center, and so it acts as the ultimate "whole system" (Fig. 10-11). The review starts in the center: a unit-length half octahedron is tacked onto each square face of the nuclear VE, thereby forming a two-frequency octahedron, which in turn has two-frequency tetrahedra added to four of its eight faces to create the large tetrahedron. Figure 10-10 illustrates this transition, and Figure 10-11 shows the complete four-frequency tetrahedron and its implied hierarchical system, employing progressively thicker lines to emphasize the three different polyhedra.

Cosmic Hierarchy (of Nuclear Event Patternings)

"The Cosmic Hierarchy is comprised of the tetrahedron's intertransformable interrelationships" (100.403b). Fuller's curious description is now clear, for we have become familiar with most of these "intertransformable relationships" and how they fit into the IVM context—as well as with the simple operations that transform one shape into another.

The order of this polyhedral hierarchy is determined by complexity, from least to most. It is worth noting that the ladder can extend *in*ward indefinitely by progressive subdivision; edges can be continually bisected to generate higher frequency systems. Notice how convenient it is to have "conceptuality independent of size." We do not have to specify the size of the initial tetrahedron, for Fuller's use of frequency to designate length provides a means to specify the system's geometric characteristics precisely, without recourse to "special case" examples. Ratios remain consistent; like conceptuality, they are independent of size.

In summation, "cosmic hierarchy" pertains to volume ratios as well as to complexity and frequency, and its relationships are uncovered through "multiplication by division." In this way, Fuller describes the order inherent in space.

Volume Reconsidered

Fuller attaches considerable significance to volume ratios and cosmic hierarchy. The subject suggests a number of philosophical implications, concerning both reasons why these orderly relationships go unnoticed and also the potential benefits of a mathematics that emphasizes systems and the relationships of parts to wholes. The first aspect is best summarized by Loeb in his "Contribution":

Uncritical acceptance of geometrical formulas as fundamental laws, particularly in systems that do not naturally fit orthogonal Cartesian coordinates, frequently leads to unnecessarily clumsy calculations and tends to obscure fundamental relationships. It is well to avoid instilling too rigid a faith in the orthogonal system into students of tender and impressionable age![3]

Overdependence on the cube is the culprit. Fuller attributes much of our attatchment to this building block to an understandable desire for "monological" solutions, or single answers to complex questions. Blissfully unaware of the "inherent complementarity" of Universe, he cautions, humanity naturally sought one "building block" with which to understand space, a single unit to be the basis of all mathematics. Even without considering the concept of "inherent complementarity," there are significant advantages to using the tetrahedron instead of the cube as the basic unit in quantifying volume. Bucky would complete his argument by reminding us that the cube is inefficient. Having demonstrated the respective volumes of this traditional shape compared to nature's minimum system, he would summarize by saying, "if you use cubes, you use three times as much space as necessary." And, once again, "Nature is always most economical."

Finally, he hypothesizes that the irrational volumes of simple polyhedra, inherent in cubic accounting, tended to reduce the importance of these basic systems in the eyes of mathematicians. Shapes with such troublesome volumes could not possibly be relevant to natural order:

> Though almost all the involved geometries were long well known, they had always been quantized in terms of the cube as volumetric unity...; this method produced such a disarray of irrational fraction values as to imply that the other polyhedra were only side-show geometric freaks or, at best, "interesting aesthetic objets d'art." (454.02)

The exclusive adoption of the cube thus served to inhibit sustained serious attention to the other polyhedra.

Back out to the big picture. To understand Universe, Fuller argued, we must think in terms of the synergetic principles governing the relationship of parts to whole systems. Multiplication by division is one of many exercises to encourage the development of a habitual orientation toward solving problems in context. If our early mathematics training encourages us to isolate and consider parts separately, rather than as components of a larger system, then, Fuller thought, our natural inclination throughout life would be to view problems myopically.

In Fuller's view, we have been blinded to a whole family of rational order by an initial (90-degree) wrong turn—itself a result of humanity's early perception of an up–down platform Earth.

11

Jitterbug

Synergetics can be described as dynamic geometry. Its treatment of polyhedra as vector diagrams and emphasis on the changes and transformations in systems distinguishes Fuller's work from the traditional geometric approach. His conviction that mathematics ought to supply dynamic models—in recognition of dynamic Universe—led to a number of interesting discoveries. "Jitterbug" is the most striking example.

Twelve equiradius spheres pack tightly around one, as noted earlier, and if the nuclear sphere is removed the other twelve can shift slightly inward. A vector equilibrium thereby contracts into the triangulated icosahedron. What would this transformation from Chapter 8 look like in terms of vectors?

Following the example set by the transition from closepacked spheres to the isotropic vector matrix, we replace spheres with vectors. Twenty-four wooden dowels and twelve four-way rubber connectors are put together to make a vector equilibrium (Fig. 11-1a). The model consists of eight triangles and six squares flexibly hinged together, and intentionally lacks the VE radial vectors, which would correspond to the now missing nuclear sphere. Therefore, since square windows collapse, we are back to the issue of stability.

In Chapter 5, we asked how many additional sticks were needed to stabilize each unstable system. For the VE, the answer was simple: six, one to brace each unstable square window. This time around we take a more open-minded approach to unstable systems and see where it leads. Suppose we *don't* stabilize the VE?

Instability enables motion. But what kind of motion? This new outlook inspires us to explore the ways in which flexible vector models change. The discoveries are remarkably satisfying (and the procedures are somewhat playful). In the case of the VE, the result is an elegant dance of symmetry that Bucky called the "jitterbug."

"Jitterbug" describes a transformation of the stick-model VE, in which all twelve vertices move toward the system's center at the

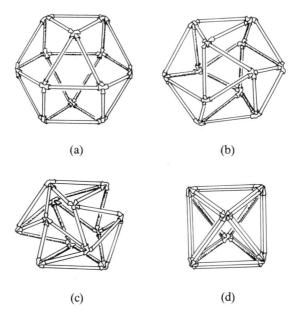

(a)   (b)

(c)   (d)

Fig. 11-1

same rate. The advantages of having an actual model on hand at this point are greater for the jitterbug than for any of the previous concepts. A verbal account of the transformation, no matter how precise, is inadequate; likewise for drawings. The wonder of the jitterbug lies in its motion—from the unique equilibrium arrangement through various disorderly stages and on to new order. Jitterbug models tend to capture the attention of the most disinterested bystander; their dance is fascinating to watch. So find twenty-four sticks and twelve connectors and put them together; the model is guaranteed to intrigue.

Folding a Polyhedron

The model fascinates because so much seems to be happening at once. The vector-equilibrium starting point looks simple enough. Unstable if left to its own devices, the VE must be deliberately held open—with one triangle flat against a table top and two hands holding the opposite (top) triangle. Notice that the two triangles point in opposite directions, together forming a six-pointed star if you peer into the system from above (Fig. 11-1a).

Now, simply lower the top triangle toward its opposite triangle (i.e., toward the table) without allowing either one to rotate. The first surprise is that the equator seems to be twisting, despite your careful avoidance of rotation. If you pull the triangle back out and try it again, you will see that this equatorial twist can go in either direction; in fact the system can oscillate back and forth, going through the zero point, or equilibrium, every time.

Secondly, although you are only pushing on one direction (forcing the triangle toward the table) the entire system contracts symmetrically—like a round balloon slowly deflating. Apparently, the effects of your unidirectional force are omnidirectional. You can see and feel that although you push and pull along a single line the contraction and expansion are both uniformly spherical.

Focus on the square windows, because only squares can change; triangles hold their shape. As the top triangle approaches the bottom triangle, each square compresses slightly, becoming a fat diamond, and the dance has begun. The radius of the system is now slightly shorter than the length of its twenty-four edges (vector equilibrium no longer). The dance continues as the top triangle approaches the bottom triangle and the diamonds grow slightly narrower, reaching the point at which their width is exactly equal to the edge length (Fig. 11-1b). The VE has thus turned into an icosahedron—at least the *shape* of an icosahedron—but we don't stop to add the six extra sticks across the diamond windows to complete the picture, for that would turn our jitterbug into a stable structure.

Instead, keep going: past the icosahedral stage, the diamonds, ever thinner, soon become narrow slits and finally snap shut (Fig. 11-1c, d). The dance comes to a halt in the form of an octahedron. Twenty-four sticks have come together into twelve pairs, creating a double-edge octahedron (Fig. 11-1d).

This contraction is continuous, and so there are countless slightly different stages, but only three are significant geometric shapes. The jitterbug is of interest primarily because of its surprising flow from one polyhedron into another; the emphasis is on its motion. However, we do want to be familiar with its geometric check-points.

First, the vector equilibrium. Unit-vector edges are balanced by unit-length radii. When the system contracts to the icosahedral position, the distances between each vertex and its five neighboring vertices are suddenly the same—unlike the VE, in which each vertex has only four nearest neighbors, each one the unit distance away, while two additional neighbors are approximately 1.414 units away,

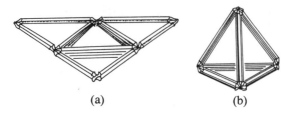

(a)   (b)

Fig. 11-2

i.e., the length of the square's diagonal. To accommodate this surface equivalence, the radius must decrease from 1.0 to 0.9511. Nature will not compromise on these numbers: for all neighboring vertices to be separated by equal lengths, the radius must be shorter than that length. A perfect static balance is impossible; hence the dynamic, eternally fluctuating events of Universe.

The twelve vertices continue their inward journey until they land at the six corners of an octahedron; sets of dowels clamp together, grouping the twelve vertices into six pairs (Fig. 11-1d). The radius has decreased to 0.7071 times its original length, and that's the end.

But wait! There's another twist left in the jitterbug. Hold on to that top triangle, which has been so carefully kept from rotating until now, and deliberately start to twist it. (The triangle will only yield in one direction, depending on the direction of the jitterbug's initial twist.) If you turn it far enough (180 degrees) the entire system collapses into a flat two-frequency triangle spread out on the table (Fig. 11-2a). Then, fold in the three corner triangles, like petals of a flower, bringing their edges together to create the fourth and final stage: the minimum system of Universe (Fig. 11-2b).

It's a dense tetrahedron, with four parallel sticks for each of its six edges, three converged vertices at each of its four corners, and a radius of 0.6124. "Quadrivalent," says Bucky, and hence full of explosive potential—ready to spring back out into "our friend the vector equilibrium."

This folded model also demonstrates the tetrahedron "turning itself inside out," for the three petals can be opened and flattened out again, and then folded back in the opposite direction, creating the mirror-image or "negative" tetrahedron. Fuller then reminds us that "unity is plural and at minimum two" and every system has an invisible negative counterpart. "Negative Universe is the complementary but invisible Universe" (351.00). Such digressions are unavoidable; for Fuller the implications of a model are always multifaceted, one observation plunging into another, layers upon layers, intertwined.

Volume and Phase Changes

The jitterbug exhibits a total transformation of shape and size without any change in material. Nothing is added or taken away, but the system's characteristics are profoundly altered by rearrangement of the parts. The concept is reminiscent of the differences between ice, water, and water vapor, all consisting exclusively of H_2O molecules. When a child—whose model-making experience is limited to "building blocks"—first learns that rearrangement of the constituents is responsible for these profound changes, the idea is not easily accepted. Fuller maintains that experience with models like the jitterbug would better prepare a child for the lessons of science. Chemistry's invisible phase changes would seem perfectly logical, as they would be consistent with first-hand experience. And indeed, the jitterbug's floppy, flexible behavior as VE is parallel to that of a gas: the dense tetrahedral configuration with its motion totally restrained is more like the "solid" phase. Same stuff, radically different properties. He has a point. These dynamic models inspire a different kind of conceptualizing.

At the zero point, twenty-four wooden dowels and twelve rubber connectors embrace a volume of twenty, as we recall from "Multiplication by Division." After contracting and twisting through the jitterbug, the bundle of sticks encloses a single unit of volume, one tetrahedron. The system has thus gone from twenty tetrahedron volumes to one, with a stop at four, in the octahedron.

Icosahedron

The icosahedron however refuses to cooperate. Its volume of approximately 18.51 is not as appealing as the whole-number ratios shared by the other stopping points. Jitterbug now introduces a rationale. The icosahedron is a phase that falls in between octahedron and vector equilibrium, rather than a definitive stopping point in the flow. The jitterbug is a continuous transformation through countless transitional stages, both regular and not, and at certain intervals an ordered polyhedron emerges. Found when the jitterbug is simply open as far as possible, the cuboctahedron is definitive, absolute zero. The octahedron clicks into place when six pairs of vertices suddenly come together. No ambiguity at either point. The icosahedral stage on the other hand is always approximate; we have to eye the distances between vertices, guessing whether or not they are equal to one. The dance does not stop naturally at this point; we just recognize the familiar shape along the way from VE to octahedron. It is thus a transient phase of the jitterbug—with no reason to stop and rest, no choice but to continue.

Similarly, the icosahedral vertices fall in between nodes of the IVM, the omnisymmetrical framework that outlines most of the symmetrical geometric shapes. One result of being out of phase with this matrix (which is also defined by closepacked spheres) is that the icosahedron's frequency cannot be increased by surrounding it with additional layers of spheres, or vectors. It cannot grow modularly; the initial choice of size, or frequency, is final. To change the frequency, a new model must be built from scratch. The icosahedron is thus restricted to single-layer construction. "The icosahedron must collapse to exist," explains Bucky; it always "behaves independently" of the other polyhedra:

> The icosahedron goes out of rational tunability due to its radius being too little to permit it having the same-size nuclear sphere, therefore putting it in a different frequency system. (461.05)

Accordingly, its volume does not fit into the "cosmic hierarchy" of rational systems.

Single Layer versus IVM

What are the consequences of the icosahedron's "independence" of the cosmic hierarchy? As a collapsed VE, it is always a shell, a single-layer construction:

> The icosahedron, in order to contract, must be a single-layer affair. You could not have two adjacent layers of vector equilibria and then have them collapse to become the icosahedron.... So you can only have this contraction in a single-layer of the vector equilibrium, and it has to be an outside layer, remote from other layers.... It may have as high a frequency as nature may require. The center is vacant. (456.20–1)

Accordingly, as we recall from Chapter 8, the design chosen by nature for many protective shells involves icosahedral symmetry—from the microscopic virus capsid to larger (visible with an ordinary microscope) radiolaria, the cornea of an eye, and a plethora of other elegant creations.

"Trans-Universe" versus "Locally Operative"

"The vector-equilibrium railroad tracks are trans-Universe, but the icosahedron is a locally operative system" (458.12). Fuller's ambiguous and somewhat mystical declaration becomes almost straightforward after the jitterbug demonstration. Vector equilibrium is incorporated into an infinitely extending network. Conceptual and timeless, VE is everywhere; it is the balance of forces at the root of all phenomena. The icosahedron, on the other hand, is always a special-case collapse, an aberration in the omnisymmetrical frame of reference. Aberrations are finite, local, inescapably stuck in time, as we recall from "Angular Topology." The icosahedron is thus fundamentally different from VE, which—with its timeless perfection—permeates all of Universe.

Fives

Fivefold symmetry dominates the icosahedron, distinguishing it once again from the cosmic hierarchy, with three-, four-, and sixfold rotational symmetries. This is another sign of the icosahedron's nonconformism. It's full of fives: to begin with, the obvious five triangles around each vertex, determining the symmetry about each of its long axes. Then, its thirty edges fall into *five* sets of six orthogonal edges, that is, three parallel pairs of mutually perpendicular edges. Figure 11-3a highlights the five distinct sets of orthogonal edges.[1] (The edges can also be grouped into sets of five parallel edges

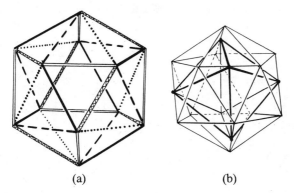

(a)   (b)

Fig. 11-3. (a) Five sets of six edges: each set of six consists of three mutually perpendicular pairs. (b) Connecting midpoints of one set of six edges outlines a regular octahedron.

embracing the equator, in six different directions.) Joining the midpoints of the six edges of one set displayed in Figure 11-3a, we discover an octahedron hiding inside—implicit in the icosahedral symmetry—in one of five possible orientations (Fig. 11-3b). The icosahedron may be out of phase with the rest of the IVM family, but it displays many significant relationships to these other shapes, which, being unexpected, are all the more fascinating to uncover. A few examples will be given below, and there's always room for further exploration. Just as in our earlier development of the cosmic hierarchy, we investigate how various shapes fit inside each other, and thereby learn about similarities in shape, volume, and valency.

Whereas the cosmic-hierarchy relationships are consistently straightforward and balanced (just bisect edges and connect midpoints to generate the next shape), whenever the icosahedron is introduced, more intricate connections emerge. We therefore have to look somewhat harder to find these new relationships which highlight the icosahedron's transitional role in the hierarchy.

We saw how the octahedron emerges out of the arrangement of icosahedral edges, on the inside, and now we reverse the situation. The icosahedron can be oriented so that eight of its twenty faces are coplanar with and flush against the eight faces of a surrounding octahedron, while the twelve icosahedral vertices are located on its twelve edges. However, the icosahedron must sit in a skew (or twisted) position, with its vertices intersecting the octahedral edges off center, dividing each edge into two segments, the longer 1.618 times the length of the shorter. This asymmetry means that there are two distinct orientations of the icosahedron inside the octahedron— positive and negative, as shown in Figure 11-4a, b.

The ratio 1.618 to 1, known as the "golden section," might have played a prominent role in synergetics, for it shows up frequently

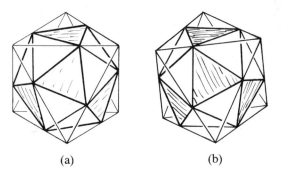

(a)   (b)

Fig. 11-4

(especially in relation to the icosahedron); however, Fuller rarely mentions this intriguing number. Accordingly, this text will not spend time exploring the famous ratio, which—as a source of fascination to geometers for millenia—enjoys considerable press already. (Specifically, for the role played by the golden section in the icosahedron, see Loeb's "Contribution to Synergetics" and its "Addendum" in *Synergetics 2*, both Section G.[2]) Fuller has a different method of coping with such relationships; rather than describing certain comparisons and their numerical values, he employs geometric "modules"—a holistic way of describing geometry with geometry.

"S-Modules"

The icosahedron's crooked position within an inscribing octahedron defines specific leftover space—six pockets of empty territory between the icosahedron and its octahedral embrace (Fig. 11-5). The symmetry of the six identical pockets is such that each can be split in half, producing two equivalent irregular tetrahedra, which then further divide into two mirror-image halves. This final thin tetrahedron is Fuller's "S-quanta module." It is a volumetric unit that describes the degree to which the icosahedron is out of phase with the IVM. Just as grade-school "long division" introduces the arithmetic remainder, the process of dividing an octahedron by an icosahedron requires a geometric remainder—the S-module. Chapter 13 will describe Fuller's A- and B-modules, volumetric counterparts of the S-module; taken altogether these quantum units comprise Fuller's finite accounting system. Finally, Figure 11-5 indicates the golden-section ratio between different edge lengths of the S-module.

Icosahedron and Rhombic Dodecahedron

A pattern emerges. The out-of-phase relationship between icosahedron and different IVM polyhedra appears to involve the golden section. We test the pattern on the rhombic dodecahedron. Do its twelve faces correspond to the twelve vertices of the icosahedron? It turns out that the two shapes exhibit an interesting relationship, with the icosahedron fitting inside the rhombic dodecahedron, predictably in a skew position. Its vertices impinge on the rhombic faces slightly off center, dividing the long diagonal of each diamond into two unequal segments—the longer again 1.618

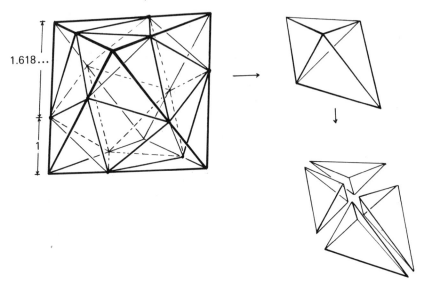

Fig. 11-5. S-module.

times the length of the shorter (Fig. 11-6). Ever reliable, the golden section reinforces our awareness of the underlying order in space.

Pentagonal Dodecahedron

Finally, recall the pentagonal faces of the icosahedron's dual; the fivefold symmetry of this dodecahedron is right out in the open. Furthermore, the pentagon is a prime source of golden section ratios (see Loeb's "Contribution to Synergetics"). The pentagonal

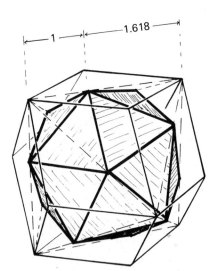

Fig. 11-6. The golden-section ratio is revealed in the relationship between icosahedron and rhombic dodecahedron.

dodecahedron is of course also out of phase with the IVM, for its symmetry scheme is the same as that of its dual, the icosahedron.

Four Dimensions

Back to the jitterbug. Fuller proposes that this fluid transition from stage to stage is best described as four-dimensional:

> The vector-equilibrium model displays four-dimensional hexagonal central cross section.... (966.04)

> Four-dimensionality evolves in omnisymmetric equality of radial and chordal rates of convergence and divergence.... (966.02)

First of all, radial and chordal equivalence produces four distinct planes, and secondly, the jitterbug contraction operates around four independent axes. Let's see how this works.

Triangles hold their shape, and therefore an equivalent model to the stick jitterbug described above can be built out of eight cardboard triangles hinged together with strong tape. The advantage in this case of "solid" triangles is that it makes Fuller's "four-dimensional" assignment easier to understand (Fig. 11-7).

The eight triangles operate in pairs. Diametrically opposite triangles remain aligned, rotating synchronously about a common axis as they together approach the jitterbug center at a constant rate. The separate pairs thus rotate around four different axes, displaying simultaneous motion in four distinct directions.

"Push straight toward the table; don't let either triangle rotate": Bucky emphasizes the simplicity of the task. He then feigns surprise at the subsequent twisting at the equator. The entire system converges symmetrically, despite the unidirectional force. He calls this behavior four-dimensionality, referring to the four independent directions of rotation.

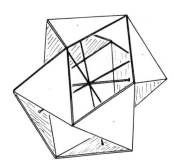

Fig. 11-7. Four independent axes of rotation.

Complex of Jitterbugs

In an isotropic vector matrix, adjacent vector equilibria create octahedral cavities; to fill space the two polyhedra must occur in equal numbers. Contemplating this alternating array in light of the VE's jitterbug behavior, Fuller suspected that a complex of jitterbugs could be synchronized to twist and contract, while their octahedral counterparts simultaneously expand, twisting open into VEs. It is an extraordinarily difficult vision to conjure up in the mind's eye: an array of synchronized twisting triangles—transforming from order to chaotic inscrutability and then back into order—but with all the places switched. VEs become octahedra; octahedra become VEs.

The ease of confirming the jitterbugging of one VE does little toward answering the question of whether a number of interconnected jitterbugs can coordinate to create this dynamic labyrinth. Would additional VEs packed around a single one serve to lock it in place? The question is difficult to answer without actually putting it to the test with the aid of a model—an awesome task. Thanks to an ingeniously designed four-valent universal joint[3] to accommodate the intricate twisting of adjacent triangular plates, a magnificent sculpture has emerged after considerable speculation. A movable complex of stainless steel and aluminum triangles hinged together effectively demonstrates that the convoluted transformation is possible (Photo. 11-1). This translation from abstract mathematical concept to physical manifestation of the motion is both an engineering and an aesthetic feat.

Fuller proposes that this complex of pulsing jitterbugs demonstrates the effects of a force propagating through space—a tangible display of otherwise invisible energy events:

1032.20 Energy Wave Propagation: ...You introduce just one energy action—push or pull—into the field, and its inertia provides the reaction to your push or pull; the resultant propagates the...omni-intertransformations whose comprehensive synergetic effect in turn propagates an omnidirectional wave. (1032.20)

In other words, the unique symmetry of the VE combines with this newfound jitterbug property to produce a model of omnisymmetrical motion, a radiating wave of activity. Just as the IVM is a static conceptual framework—describing the symmetry of space—this model illustrates the concept of dynamic, "eternally pulsating" energy events in space. It causes the IVM to come to life.

A model can elucidate a concept without being an exact duplicate of the phenomenon in question. In fact, considering the oddly

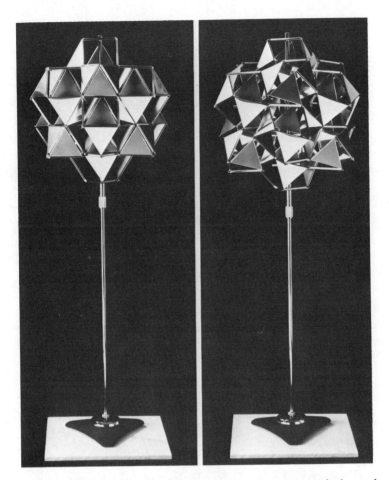

Photo. 11-1. Complex of jitterbugs: An array of alternating octahedra and vector equilibria is shown on the left; on the right, the display is undergoing the simultaneous transformation of all cells, such that octahedra are opening up to become VEs while the VEs are contracting into octahedra. The action is frozen in mid transformation, making it possible to see the icosahedral phase. Photo courtesy of Carl Solway Gallery, Cinncinnati, Ohio.

mystical language that creeps into modern physicists' description of atomic and subatomic behavior, we can conclude that invisible reality does not readily submit to large-scale reenactment with "solid" materials. The intention of Fuller's models therefore is to provide a consistent analogy—a tangible display that parallels and thereby explains invisible behavior:

Dropping a stone in the water discloses a planar pattern of precessional wave regeneration. The local unit-energy force articulates an omnidirectional, spherically expanding, four-dimensional counterpart of the planar water waves' circular expansion. (1032.20)

The expanding concentric waves made by a stone dropped in a lake are directly visible on the water's surface. It is therefore easy to picture the image of a wave propagating across a plane. It is more difficult to visualize a corresponding situation in space, which is precisely the territory Fuller set out to conquer. He strove to clarify invisible aspects of reality through models that can be seen and felt. The complex of jitterbugs makes the concept of an expanding spherical wave of energy visible.

Whatever the analogous events in Universe, the model is intricate and phenomenal. Fuller's argument is that nature depends upon such dynamic orderly coordination. The complex of jitterbugs demonstrates a complicated but organized operation, and if nature permits this transformation of her omnisymmetrical framework, she might use the same trick elsewhere. At the moment, jitterbugs therefore merely hint at possibilities. It is worth reflecting on the extraordinary intuition required to have discovered this subtle and magnificent geometric phenomenon.

Other Dynamic Models

Topology and Phase

Physical Universe differentiates into three categories: liquid, crystalline, and gaseous phases. These distinct states of matter arise as a result of changing temperature or pressure, which induce different types of bonds. Fuller proposes a simple geometric analogy to make these invisible changes easy to comprehend. Appropriately, the model is completed with an exhaustive enumeration of the ways in which two tetrahedra can be connected to each other. The three arrangements of the minimum conceptual system model the three phases of physical matter.

First, two tetrahedra are triple-bonded, sharing one face between them (Fig. 11-8a). Because the relative positions of the two tetrahedra are completely fixed, the arrangement qualifies as a stable set of relationships and represents a crystalline structure: "The closest-packing, triple-bonded, fixed-end arrangement corresponds with rigid-structure molecular compounds" (931.60). Once again, the intention is not to create a large-scale duplicate of a particular "solid" compound, but rather to display the different characteristics of each chemical phase. The model is a kind of visual shorthand.

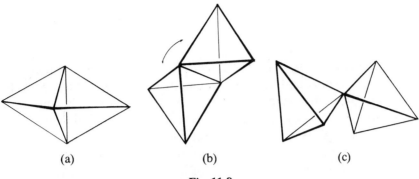

(a)   (b)   (c)

Fig. 11-8

One of the three bonds is now released, leaving a double bond between two tetrahedra. The shared edge acts like a hinge (Fig. 11-8b); the tetrahedral pair can swing back and forth, but they cannot be moved closer together or farther apart. The configuration is thus noncompressible—one of the distinguishing characteristics of liquids. It bends any way you desire—malleable just like a liquid—but the double bond persists:

> The medium-packed condition of a double-bonded, hinged arrangement is still flexible, but sum-totally as an aggregate, all space-filling complex is noncompressible—as are liquids. (931.60)

Finally, we break another of the bonds. "Single-bonded" tetrahedra are joined by a vertex, and their behavior is analogous to that of a gas. The vertex bond acts as a universal joint; the two halves can swing freely with respect to each other, moving together and apart without disrupting the type of bond (Fig. 11-8c). The arrangement is compressible, expandable, and completely flexible—short of dissociation. Perpetual connectedness indicates that both tetrahedra continue to participate in the same substance; they exhibit a consistent relationship, but lack structural definition:

> Tetrahedra linked together entirely by...single-bonded universal jointing use lots of space, which is the openmost condition of flexibility and mutability characterizing the behavior of gases. (931.60)

The analogy is complete. All the while, Bucky holds the simple structures in his hands, and explains the different basic properties of solids, liquids, and gases. With any luck, a small child is present, forcing him to keep his discourse simple. How can the same type of molecule produce such radically different substances? Bucky offers a tangible explanation through geometry. This model is perhaps useful as a mnemonic device—an easy way to remember the chemistry lesson by relating the different characteristics to their analogous stage of the model—rather than as a true demonstration of phase changes in a substance.

It is worth noting that this model of interconnected tetrahedra is more appropriate, and, in fact, quite accurate, in connection with the bonding of carbon atoms within molecules. In conclusion, it is unfortunate that this and other polyhedral characteristics and relationships are generally overlooked in educational curricula.

Fuller developed many dynamic models, and readers who go on to further study will find a variety of examples in *Synergetics*. The transformations discussed in this chapter set the stage to explore other examples. Appropriate parallels in nature may well arise from such efforts.

"All-Space" Filling:
New Types of Packing Crates

Fuller in his characteristic drive for verbal accuracy updates geometry's conventional term "space filling" with his own more descriptive (and predictably longer) "all-space filling." Here's the puzzle: which of the polyhedra introduced so far can pack together in such a way that all available volume is occupied without any gaps?

The concept is not new; ever since closepacking equiradius spheres, we have danced around the issue of space filling, and in the process made most of the discoveries that this chapter will expand upon. The IVM disclosed certain space fillers, while making it clear that other polyhedra did not share this ability. However, filling all space now becomes the focus of our investigation, calling for the systematic analysis that enables new insights and a more thorough understanding.

Despite its obvious applicability, space filling is not emphasized in Fuller's work. An ability to "fill all space" is generally mentioned as further description of a given polyhedron rather than providing an investigative starting point for synergetics. Fuller's "operational" approach encourages more experimental exercises, such as packing spheres together, which then lead to space fillers after the fact.

In view of our overall goal of researching the characteristics of space, what could be more logical than to ask what fits into it? What shapes are accommodated by space? The notion that space is not a passive vacuum gradually becomes second nature; experience has changed our awareness. Now we want to become ever more exact about these active properties. The existence of an extremely limited group of polyhedra that can pack together to fill all space is one of the more direct illustrations of the specificity of spatial characteristics.

The puzzle is quite challenging. Without actually making a horde of tiny cardboard models of the polyhedra in question, these spatial configurations are extraordinarily difficult to visualize—with the sole exception of the obvious space filler, an array of cubes. Once again, to gain more experience with the concept, we revert to the plane, or to be more accurate, the page. In deference to Bucky's strict precision, we must acknowledge that the theoretical "plane" is a nondemonstrable concept, but we can certainly (and quite' appropriately) discuss filling up a page.

Plane Tessellations

The mathematical title may be somewhat intimidating, but plane tessellations are actually quite familiar. Derived from the Latin word for "tiling," tessellation, in mathematics, refers to planar patterns of polygons. Because of the ease of working with flat patterns, we begin our study of space filling with the analogous situation in the plane. Which regular polygons fit together edge to edge to fill a page?

We can fill up a page with squares, as anyone who has ever seen graph paper knows, and the pattern created by equilateral triangles is almost as familiar. Another successful tessellation is found on many bathroom floors covered by tiny hexagonal tiles (Fig. 12-1a). With these immediately apparent examples, we might begin to suspect that any regular polygon can fill a page. However, a little experimentation quickly reveals that we have already exhausted the possibilities.

Three regular pentagons (equilateral and equiangular) placed side by side leave a 36-degree angular gap—not nearly wide enough to accommodate a fourth pentagon. Five-sided tiles are thus disqualified. Three heptagons simply cannot fit around a single point, as we saw in Chapter 4; octagons meet a similar fate, as of course do any polygons with more sides. We are suddenly confronted with a very limited group of plane fillers: triangles, squares, and hexagons.

Opening up the field to allow combinations of regular polygons while maintaining equivalent vertices expands the inventory only slightly. Eight "semiregular" tessellations join the three patterns above (Fig. 12-1b). Still an impressively small group. These eleven tilings can be categorized in terms of the different rotational symmetries exhibited by each, and within this group every category of repeating planar pattern is represented. It is fascinating to reflect on the implications of these results: for example, a wallpaper designer can only create what the limitations inherent in the plane will allow.

These cumulative experiences—especially those as straightforward as plane tessellations—nurture our growing awareness of spatial

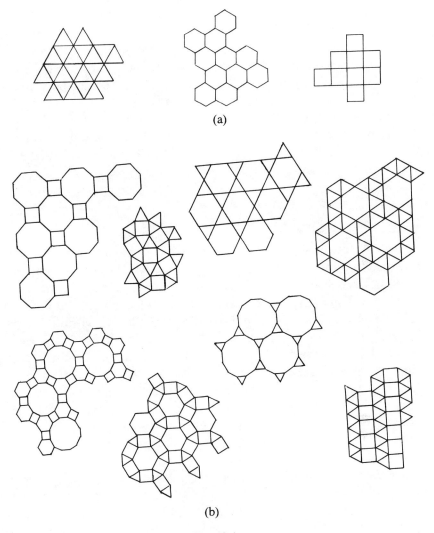

(a)

(b)

Fig. 12-1

constraints. And with each step, our knowledge of the elegant precision of this order expands synergetically.

Filling Space

Cubes stack neatly together to fill space. What other polyhedra exhibit this property? Attempts to fit regular tetrahedra together are quickly frustrated; likewise for octahedra. However, working together, the two shapes can fill space indefinitely. None of this is new information: we discovered the complementarity of octahedra and tetrahedra while exploring isometric arrays of both spheres and vectors. Subsequently, multiplication by division uncovered the octahedron hiding inside every tetrahedron. The octahedron–tetrahedron marriage is clearly an eternal bond.

Neither icosahedra nor pentagonal dodecahedra can fill all space. ["Icosahedra, though symmetrical in themselves, will not close-pack with one another or with any other symmetrical polyhedra" (910.01).] The cube thus stands alone among regular polyhedra.

Complementarity

But let's reevaluate our apparently simple array of cubes. As the obvious solution to filling all space with a single polyhedron, this packing seems to provide the most straightforward information about the shape of space. But look further. What if you could see the cubes' face diagonals? An implied tetrahedron awaits visibility. Now imagine filling in the necessary diagonals, so that the inscribed tetrahedron—surrounded by four eighth-octahedra—appears in each cube. At every junction of eight cubes, the octahedral parts come together and form one complete octahedron around each cubical corner (Fig. 12-2). As rectilinear boxes are unstable without diagonal bracing, a stabilized packing of cubes turns into an Octet Truss. Whether visible or not, the octet symmetry is implicit in the configuration.

The Greeks failed to get at the triangulated heart of their stack of cubes, philosophizes Bucky, for "like all humans they were innately intent upon finding *the* 'Building Block' of Universe." Had they experimented with arranging tetrahedra vertex to vertex and been

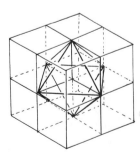

Fig. 12-2. Octahedron at junction of eight cubes.

confronted with the inescapable octahedral cavities, he continues, they

would have anticipated the physicists' 1922 discovery of "fundamental complementarity.". . .But the Greeks did not do so, and they tied up humanity's accounting with the cube which now, two thousand years later, has humanity in a lethal bind of 99 percent scientific illiteracy. (986.049b)

Fueled by the developments of twentieth-century physics, Fuller spoke frequently and emphatically of the "inherent complementarity" of Universe. He cites two examples in particular out of the many provided by quantum physics. First, the "complementarity principle" announced by Niels Bohr (1885–1962) in 1928, which goes hand in hand with Heisenberg's indeterminism. Bohr summarizes the basic feature of quantum physics by stating that experimental evidence cannot be comprehended within a single frame, but rather must be understood as "complementary," or partial, information. The totality of a phenomenon must therefore be represented through more than one complementary part, for all aspects cannot be accurately measured simultaneously.[1]

Fuller also calls our attention to a later Nobel-winning development made by two Chinese physicists working in the U.S. In 1957, Tsung Dao Lee and Chen Ning Yang were honored for their discovery that "parity" is not conserved in weak interactions, for the subatomic particles involved show "handedness."[2] In other words, simplifies Bucky, fundamental complementarity does not consist of mirror-image pairs. Concave–convex, tension–compression, proton–neutron, male–female: Universe is always plural, supporting interdependent, inseparable pairs which are not simply mirror-image halves. It is thus reassuringly consistent that the essence of structure and space should also exhibit this fundamental dualism. There *is* no single building block of Universe.

Other Space Fillers

The role played by tetrahedra and octahedra in an array of cubes—not to mention in the IVM—demonstrates that these two structures combine to create a variety of polyhedra. This observation suggests an operational strategy: experiment with various combinations in the hope of finding other space fillers.

Two tetrahedra are affixed to opposite faces of an octahedron with the same edge length. Supplementary dihedral angles cause adjacent triangular faces to be coplanar, thereby merging into six rhombic faces (Fig. 9-8). The resulting slanted structure, introduced in Chapter 9, is called a rhombohedron and can be thought of as a partially flattened cube. Six squares have simply been squashed into diamonds. It's not hard to imagine that an entire array of toothpick cubes could lean over in unison, transforming into an array of distorted cubes. Rhombohedra therefore fill space. The simplicity of this development hints at a starting point.

In his "Contribution to Synergetics," Loeb analyzes various polyhedra for their divisibility into tetrahedra and octahedra, and demonstrates that a shape will fill space *if it consists of two tetrahedra for every one octahedron.*[3] Suddenly out of randomness an order emerges, and trial-and-error is replaced by a generalized principle. The ability to fill all space can be added to the list of descriptive properties (such as stability, symmetry, duality) by which we categorize the scattered cast of polyhedral characters. And now, armed with Loeb's conclusive analysis, we will continue to explore the puzzle from a slightly different frame of reference.

The Search Continues

The IVM simplifies our task, providing a frame of reference which itself fills space. All of the polyhedra covered so far, with the exception of the icosahedron and pentagonal dodecahedron, are outlined within the IVM and IVM' framework by various combinations of octahedral and tetrahedral components (both parts and wholes). We can therefore survey the matrix to ascertain which of these systems are able to meet face to face without intervening cavities.

Polyhedra that fit together without gaps must completely surround a common vertex. IVM vertices therefore provide a good starting point; we can systematically investigate the different types of nodes in the matrix, checking the surrounding cells for space-filling polyhedra. Our first node reveals an already familiar space-filling team: each IVM vertex joins six octahedra and eight tetrahedra—as well as polyhedral combinations of the two. The most obvious of these combinations is the vector equilibrium, and our study of the IVM already made it clear that VEs do not fill space, but rather must cooperate with octahedra to create an uninterupted array, providing yet another example of complementarity. The necessity of this pairing follows directly from octahedron–tetrahedron interdependence. (Chapter 13 will elaborate on such space-filling teams, which arise out of the two-to-one ratio mentioned above.)

Our next step is to investigate other vertices. Interconnecting the centers of octahedra, we trace the array of minimum cubes. That one's easy. The fact that cubes fill space is not new; what else can we learn?

If octahedron centers provide the meeting point for eight cubes, what happens at the centers of tetrahedra? To begin with we observe that four quarter tetrahedra convene. The base of each shallow pyramid is the face of a neighboring octahedron, so we ask what shape is created by an octahedron framed by eight quarter tetrahedra. As seen in Chapter 9, this is equivalent to Loeb's "degenerate stellation"; the thin triangular faces of adjacent quarter tetrahedra become coplanar when surrounding an octahedron, and thereby merge into one diamond face. The result: a rhombic dodecahedron (Fig. 9-12). In IVM context, this means that every octahedron reaches out, incorporating eight neighboring quarter tetrahedra, so that the tetrahedra are completely used up (no leftovers). The entire matrix is thus involved, meaning of course that rhombic dodecahedra fill space. By interconnecting the centers of every tetrahedron in the IVM we automatically generate an array of rhombic dodecahedra.

The space-filling puzzle becomes more enticing with this new addition. With its many diamond faces and irregular surface angles, this shape is not, like the cube, an obvious space filler. A casual observer would not suspect this intricate polyhedron of fitting so beautifully together. In fact, without the advantage of the IVM, it is difficult to picture how rhombic dodecahedra manage to fill space.

The Dual Perspective

Recalling from Chapter 4 that the rhombic dodecahedron is also a "degenerately stellated" cube, it is interesting to observe the relationship between the two packings. A cube with four body diagonals dividing the inside into six pyramids is shown in Figure 12-3a. These square-based pyramids are the exact shape required to degenerately stellate a second cube. We can "unwrap" that subdivided cube and place its components on the six faces of a second intact cube—only to arrive once again at our diamond-faceted friend (Fig. 12-3b). We therefore have a new way to visualize the rhombic-dodecahedron

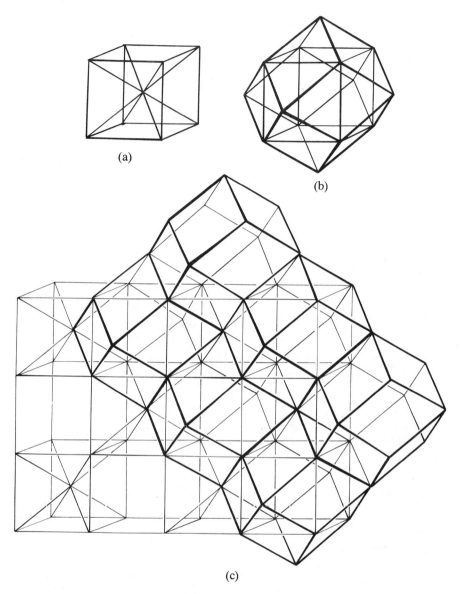

(a)

(b)

(c)

Fig. 12-3

packing. Start with an array of cubes, in which every other cube is subdivided by a central node, while the rest remain empty. We have just described an array of rhombic dodecahedra (Fig. 12-3c). A framework of cubes is so familiar and readily imagined that this exercise brings the rhombic dodecahedron's space-filling capability into easy grasp.[4]

Duality and Domain in Sphere Packing

Imagine that our closest-packed spheres are actually perfectly round balloons. Supposing they are all packed tightly into a closed container which insures that their positions are fixed, we then try to picture what would happen if more air were steadily pumped into each balloon. Remember that they are unable to move—preventing the natural reaction of collectively spreading out and taking up more space. Instead each individual balloon expands and presses more tightly against its neighbors so that the *points* of tangency merge into *planes* of tangency:

> A bubble is only a spherical bubble by itself. The minute you get two bubbles together, they develop a plane between them. (536.44)

We allow the balloons to grow to the extent that all available space is occupied. What is the shape of the balloons once they merge together? Remember that twelve spheres pack tightly around one, and so the tangency points between spheres, which were located at the twelve (four-valent) vertices of vector equilibria, must now be replaced by the same number of four-valent faces. We are thus reminded that the domain of a sphere in closest packing—in Fuller's words "the sphere and the sphere's own space"—is a rhombic dodecahedron.

The above sequence provides an important insight into the space-filling ability of rhombic dodecahedra. By renaming this diamond-faceted polyhedron "spheric," Fuller places considerable emphasis on its relationship to closest packing. The spheric is thus presented as a cosmically significant shape, the domain of the generalized energy event, and consequently, the domain of every intersection in the omnisymmetrical vector matrix.

Synergetics thus arrives at its all-space fillers through investigation of nature's omnisymmetrical framework. IVM provides the context:

> A "spheric" is any one of the rhombic dodecahedra symmetrically recurrent throughout an isotropic vector-matrix-geometry... (426.10)

> 426.20 Allspace Filling:... Each rhombic dodecahedron defines exactly the unique and omnisimilar domain of every radiantly alternate vertex... as well as the unique and omnisimilar domains... of any aggregate of closest-packed uniradius spheres..."

In a later section of *Synergetics*, Fuller expands upon the above observations with a more general statement about any point in space.

The most complete description of the domain of a point is not a vector equilibrium but a rhombic dodecahedron, because it would have to be allspace filling and because it has the most omnidirectional symmetry. The nearest thing you could get to a sphere in relation to a point, and which would fill all space, is a rhombic dodecahedron. (536.43)

Truncated Octahedron

The "truncated octahedron," or tetrakaidecahedron, is also a known space filler (Fig. 12-4a). A model of this semiregular polyhedron can be constructed by taping together eight cardboard hexagons and six squares of the same edge length. But Fuller is wary of that approach, a geometry which supports the illusion of "solids." Instead, synergetics prescribes that we view the system in context. Therefore, to generate this shape we start with a "three-frequency octahedron." With its edges divided into three equal segments, this octahedron submits quite naturally to vertex truncation, as shown in Chapter 9, Figure 9-9b. A (single-frequency) half octahedron can be chopped off from each corner, creating six square faces and converting the three-frequency triangles into equilateral hexagons.

How do they fit together? Hexagonal faces of two truncated octahedra come together after rotating 60 degrees with respect to each other, so that square faces alternate (rather than landing next to

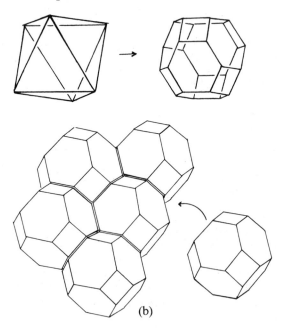

(b)

Fig. 12-4

each other) and begin to frame a cavity in the exact tetrakaideca-hedral shape (Fig. 12-4b).

Four tetrakaidecahedra fit together around one IVM vertex. To ascertain the exact relationship of this packing to the IVM will require some investigation: how remote are the vertices involved, and how many cells are incorporated in between? We will answer these questions below, but first we go back to the simpler cases, for a sense of the whole progression.

Two to One: A Review

We can now appreciate Loeb's development of the requirement that a space-filling polyhedron can be broken down into twice as many octahedra as tetrahedra.[3]

The cube consists of one tetrahedron plus four eighth octahedra—or a total of one half octahedron. This ratio of 1 to $\frac{1}{2}$ certainly qualifies. Eight quarter tetrahedra (for a total of *two*) embrace *one* octahedron to create the rhombic dodecahedron. With this confirmation, we start a list of results, displayed in Table VI.

The VE consists of eight tetrahedra and six half octahedra: 8 to 3. Lacking one octahedron, the VE must not be a space filler, a conclusion that is consistent with our earlier observation of the IVM. However, pair a VE with the missing octahedron and all space can be filled. The two-to-one ratio also serves as a prescription for complementary pairing; if an IVM system is not a space filler, the ratio tells us what's missing.

The rest? All-space-filling rhombohedra are utterly straightfor-ward; with two tetrahedra on either side of an octa, this shape gave

Table VI. Octahedron–tetrahedron ratio in space filling

Polyhedron	Ratio (tetrahedra : octahedra)
Tetrahedron	$1:0$
Octahedron	$0:1$
Cube	$1:\frac{1}{2}$ [a]
Rhombic dodecahedron	$2:1$ [a]
VE	$8:3$
Rhombohedron	$2:1$ [a]
$3v$ octahedron	$32:19$
Truncated octahedron	$32:16$ [a]
$3v$ tetrahedron	$11:4$
Truncated tetrahedron	$7:4$

[a] Space filler.

us our starting point for the space-filling ratio. Icosahedra cannot be broken down into tetrahedra or octahedra; the icosahedron is eternally out of phase. That brings us to the truncations, which can be analyzed in terms of the IVM. As a representative example, we explore the truncated octahedron.

The three-frequency octahedron incorporates nineteen single-frequency octahedra and thirty-two tetrahedra. It can be a frustrating experience to count these individual cells, but it can be done! Skeptics are encouraged to get out a box of toothpicks and some mini-marshmallows. Figure 12-5 illustrates the dissection of one half of a three-frequency octahedron, to make it easier to count the unit cells in individual layers. Sixteen tetrahedra and nineteen half octahedra are clearly visible in the three layers shown. These results are simply doubled to verify the totals in the first sentence of this paragraph.

The ratio of 32 : 19 does not qualify. Not surprisingly, the three-frequency octahedron—like any regular octahedron, for size does not affect shape—is not a space filler. The truncation process removes half of a small octahedron from each of the six corners, or three octahedra altogether. Nineteen minus three leaves sixteen, while the total of thirty-two tetrahedra does not change. 32 : 16 is indeed two to one, and the rule holds true once again.

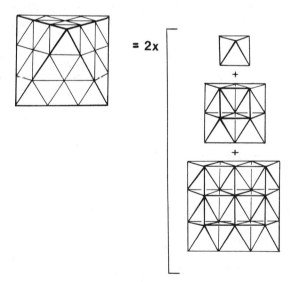

Fig. 12-5. Individual layers of three-frequency octahedron.

The truncated tetrahedron can be dissected in the same manner as its octahedral counterpart. The IVM starting point, a three-frequency tetrahedron, consists of eleven unit tetrahedra and four octahedra, and is clearly not a space filler—as we well know. Figure 12-6 enables the unit cells to be counted, by separating the layers of a four-frequency tetrahedron. In the process of counting, we can also finally verify the claim introduced in Chapter 2 that third-power numerical values (2^3, 3^3, 4^3, etc.) are represented by the volumes of tetrahedra of increasing frequency. Recall that the expression "x cubed" is derived from the fact that cubes of progressively higher frequency consist of $1, 8, 27, 64, \ldots$ unit cubes. Fuller points out that exactly the same volumetric increase is exhibited by tetrahedra, as

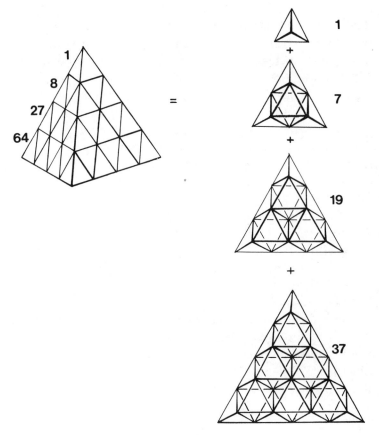

Fig. 12-6. Volume accounting in tetrahedra of increasing frequency: third-power model

portrayed in Figure 2-1b. We now take advantage of the disected four-frequency tetrahedron to verify these values. Adding up the volumes of the unit tetrahedra (each with a volume of one) and unit octahedra (each with a volume of four) in successive layers, we quickly find that the "cubic" numbers are confirmed, as shown in Figure 12-6. A two-frequency tetrahedron has a volume of 8; a three-frequency, 27; a four-frequency, 64; and so on. "Nature is not 'cubing,' she is 'tetrahedroning,'" announces the triumphant Bucky.

Back to the space-filling investigation, our next candidate is the (semiregular) truncated tetrahedron, which in synergetics is derived from a three-frequency tetrahedron. Subtracting a unit tetrahedron from each of the four corners yields a truncated tetrahedron consisting of seven small tetrahedra and four octahedra (Fig. 9-9a). Seven-to-four does not correspond to the desired ratio, and we thus learn that this shape will not fill space either; it lacks one unit tetrahedron. As a final illustration of the disection method, we note the new totals when a three-frequency tetrahedron is paired with a three-frequency octahedron. The resulting inventory consists of 43 tetrahedra and 23 octahedra, which does not quite reach the desired double ratio. What if we add a second three-frequency tetrahedron? The final total is then 54 tetrahedra and 27 octahedra, and that's it. Two $3v$ tetrahedra paired with one $3v$ octahedron are able to fill space. Once again, shape proves independent of size.

The above examples provide first-hand confirmation of the space-filling hypothesis, allowing us to feel confident that a new truth has been revealed.

Out of all the accumulated data, a consistent finding emerges. With this generalized principle, we are equipped to determine whether any polyhedron that submits to analysis in terms of octahedral and tetrahedral components is a space filler. We just have to break it down into these constituent parts and check the ratio for the requisite two-to-one. The ratio replaces the arduous task of trying to push shapes together and twist them around to see if they pack; it is another shortcut. IVM provides the framework; two-to-one is the ratio; all criteria supplied.

Coincidence? Magic? Or rules of spatial order?

The Heart of the Matter: A- and B-Quanta Modules

Minimum system, triangulation, equilibrium of vectors, closest-packed spheres, and space-filling: the path toward the isotropic vector matrix has many origins, any of which can yield the unique omnisymmetrical array. The result of traveling all of these routes is a powerful awareness of the interplay of octahedral and tetrahedral symmetries in space. However, the more thoroughly we search the IVM, the greater the intricacy of these "intertransformabilities"—calling for a new level of analysis to keep track of our discoveries.

IVM got us down to the basics. Even the cube, mathematics' conventional building-block, reduces to octahedron–tetrahedron components. However, tetrahedra and octahedra are not true structural quanta, for it was often necessary to break them apart into subcomponents in order to build other polyhedra. Our task is therefore still unfinished. We have yet to get to the heart of the tetrahedron and octahedron.

Necessity thus leads us to Fuller's A- and B-quanta modules.[1]

The missing element in our IVM analysis is a way to handle redundancy. Symmetry—the degree to which a system looks exactly the same in different orientations—is a kind of structural redundancy. The tetrahedron, the octahedron, and their various combinations all have a high degree of symmetry, and now we intend to get to the root of it. How? By subdividing symmetrically until we can go no further. As long as a system exhibits some degree of symmetry, it can be divided into identical subunits, which can be put together to recreate the original system. Ergo, the system was redundant. Through progressive subdivision, we can locate the minimum subunit of any system. This final asymmetrical module—or "least common denominator" (LCD)—contains the geometrical data needed to reconstruct the whole system. We find the LCD by subdividing a polyhedron until we reach the limit case, that is, a module that can no longer be split into similar units.

A-Quanta Modules

Let's start again with our highly symmetrical friend, the tetrahedron. The fact that its four faces are equivalent presents the first opportunity for subdivision—resulting in four equal parts. Each quarter tetrahedron encompasses the region from the center of gravity (cg) to a face (Fig. 13-1a). It is evident from the threefold symmetry of these shallow pyramids that each can be sliced into three identical pieces, as if it were a triangular pie (Fig. 13-1b). The resulting pie pieces— long thin tetrahedra stretching from the apex of the quarter tetrahedron (cg of regular tetrahedron) out to an original unit-length edge —are quite irregular, and so the process is almost complete. However, one type of symmetry conspicuously remains. Each sliver can be split in half to produce two mirror-image parts: a right- and a left-handed version with identical angles and lengths (Fig. 13-1c). And suddenly we have come to the end. There is no possible way to divide that final product into equal parts; the shape thus generated is the limit case. Fuller calls this asymmetrical tetrahedron the "A-quanta module."

A-quanta modules contain the complete geometric ingredients needed to create a regular tetrahedron. We need twenty-four A-modules (twelve positive and twelve negative) to make a tetrahedron, but one module alone supplies the information. The A-module, representing the volumetric essence of the tetrahedron, introduces a new kind of building block.

B-Quanta Modules

Having ascertained the minimum unit of the minimum system, we must not forget about inherent complementarity. The above analysis is repeated. A regular octahedron splits into eight equivalent pyramids (octants) (each of which divides into six equal pieces), as did the quarter tetrahedra, yielding forty-eight asymmetrical minimum units (LCDs) of the octahedron (Fig. 13-2a). That would have been the end of the story except for a subtle catch, an overlooked redundancy.

The eighth-octahedron has twice the altitude of the quarter tetrahedron, but both pyramids have the same (equilateral triangle) base. The quarter tetrahedron therefore fits right inside the octant, occupying exactly half the available volume (Fig. 13-2b). The *remaining* volume is pure octahedron—a hat-shaped wedge that accounts for the shape difference between the two pyramids. This concave trian-

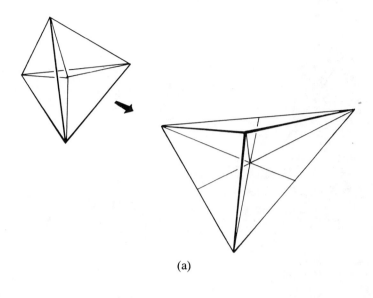

(a)

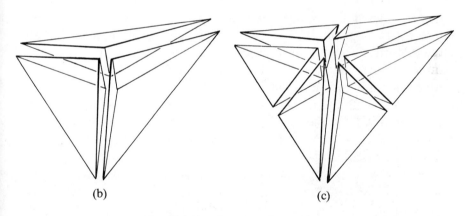

(b)   (c)

Fig. 13-1. Development of *A* module.

gular lid can be subdivided into six equal irregular tetrahedra (three positive, three negative) and these are called B-quanta modules. They are generated by the LCD of the octahedron after that of the tetrahedron is taken away. As long as the octahedron's asymmetrical unit contained a complete A-module within its boundaries, the unit was redundant. By removing the A-module which had occupied half

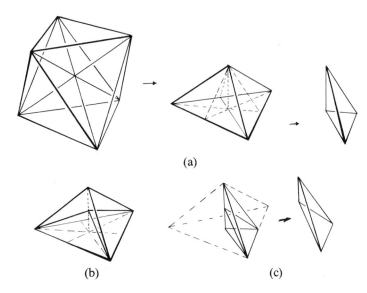

(a)

(b)   (c)

Fig. 13-2. (a) One-sixth of an "octant" equals $\frac{1}{48}$ of an octahedron. (b) The quarter tetrahedron fits inside the octant and has half the altitude. (c) Subtracting an A module from $\frac{1}{48}$ of an octahedron defines the B module.

the volume of the octahedron's LCD, we finally achieve a second true modular quantum (Fig. 13-2c). Thinner and more pointed than the A's, the B-quanta have a very different shape but precisely the same volume. Neither A nor B can be made from the other; they are fundamentally distinct, complementary equivolume modules.

The LCD of the octahedron ($\frac{1}{48}$) consists of one A and one B, while the LCD of the tetrahedron ($\frac{1}{24}$) is simply an A-module. The analysis is complete: we have broken down our IVM constituents into their essential characteristics. We can go no further.

However, the field is now wide open for experimentation. Armed with the IVM quantum units, we can anticipate a tremendous range of combinations and permutations, or rearrangements. Having systematically analyzed our basic systems, we now start to put their essential quanta back together in order to further understand the relationships between polyhedra. We have thus developed a far more sophisticated (and specific) framework with which to explore polyhedral intertransformability. Fuller takes it a step further. (Why stop with polyhedra?)

If you are willing to go along with the physicists, recognizing complementarity, then you will see that tetrahedra plus octahedra—and their common constituents, the unit-volume A- and B-Quanta Modules—provide a satisfactory way for both physical and metaphysical, generalized cosmic accounting of all human experience. (950.34)

We also observe considerable multiplication of complexity with the new framework created by subdivision. As Bucky would have it: multiplication by division.

Energy Characteristics

Progressive subdivision has left us with legitimate geometric quanta: final packages which cannot be split into equal halves. In a general sense, the model parallels science's search for the ultimate aspects of reality. Physics probes deeper and deeper into matter, breaking it down into ever-smaller constituents—cells, molecules, atoms, sub-atomic particles—ultimately seeking a package of energy (quarks?) which cannot be split apart. Fuller probes similarly into his geometry, hoping to gain insights about Universe itself.

Fuller's profound faith in the significance of reliable patterns—coupled with his unflagging determination to find nature's coordinate system—led him to draw many parallels between synergetics and nature. Most of these are both suggestive and unde-veloped; he planted his seeds, left the cultivation for posterity, and went on with his search. Certain that science can be modeled, Fuller felt a great responsibility to pursue what he saw as an ever more relevant investigation; if patterns are to emerge, sufficient data must be collected. Toward the goal of clarifying the patterns observed by Fuller, our strategy might be to forge ahead: cover as much ground as possible and worry about significance later. However, in his coverage of intriguing geometric properties, Fuller often immediately assigns connections to physical phenomena. We can neither ignore nor confirm such speculation, and so for now, we merely record and file away. The respective energy "valving" (ability to direct, store, control) properties of A- and B-modules is a typical example.

Once again, the issue is based on "operational" procedure. Models of Fuller's two tetrahedral quanta are constructed out of paper. The process starts with a planar "net" (a flat pattern piece which can be folded up and taped together to make a specific polyhedron). Con-sider for example the regular tetrahedron. Its four equilateral-trian-gle faces can be generated by folding one (double-size) triangle along lines that connect mid-edge points (Fig. 13-3a). This ability to form a tetrahedron by folding *one* triangle is not to be taken for granted; the situation that allows all four faces of a generic tetrahedron to fit

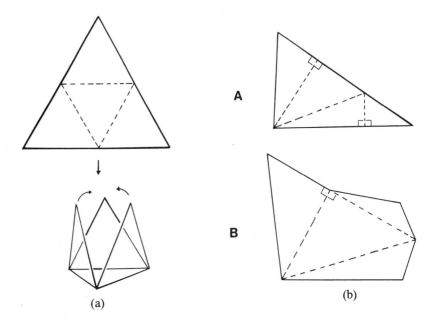

(a)   (b)

Fig. 13-3

inside a triangular frame—requiring exactly supplementary angles out of the infinite possibilities—is an exception. We are not surprised by Figure 13-3a for we expect such exceptional cooperation from the uniquely symmetrical regular tetrahedron.

The surprise is that the asymmetrical A-module unfolds into one planar triangle: "an asymmetrical triangle with three different edge sizes, yet with the rare property of folding up into a whole irregular tetrahedron" (914.01). This unusual property makes it a kind of pure form. The B-module, on the other hand, will always fold out into four separate planar triangles no matter which vertex you start with; its net will never fit into one triangular frame. Figure 13-3b compares the two nets.

Fuller connects this geometric property with "energy." A-modules are thus said to concentrate or hold energy, while B's release or distribute. This conclusion is based on the fact that "energy bounces around in A's working toward the narrowest vertex," only able to escape at a "twist vertex exit" (921.15). Comparing pattern pieces, A's planar net offers three escapes; the jagged four-triangle complex of B offers twice that number (Fig. 13-3b). It is a somewhat bizarre observation to begin with, and characteristically it led to the following ambiguous assignment of meaning: a vertex, or non-180-degree junction, in a planar net represents disorderly energy-escaping properties. Hence in A-modules, energy is seen as contained, able to bounce around inside the net without many available exits, whereas in B-modules, energy is quickly released.

We can conceive of this energy in many ways—as light beams, as bouncing electrons, or even as billiard balls for a more tangible image. All three qualify as "energy events," and having a specific image in mind makes it easier to think about the different "energy-holding" characteristics. To understand and evaluate Fuller's assertion, we go along with his use of "energy," for the word covers a great deal of territory already and his usage is internally consistent. Fuller calls our attention to a geometric property that we may not have otherwise noticed, and with respect to this phenomenon of planar nets there is no doubt as to the difference between A- and B-modules. What *is* the significance of this distinction? What are we to conclude about the orderly contained A versus the disorderly sprawling B? In terms of physical Universe, a judgement probably cannot be made. However, for the purposes of this text and of continuing to explore the geometric interactions of the two quanta, we adopt Fuller's energy assignment: A's conserve; B's dissipate. It provides a consistent reference system with which to classify the two quanta and their subsequent interactions. Furthermore, two basic modules exhibiting the same volume and different energy characteristics provide an even more attractive model of "fundamental complementarity": equivalent weight or importance, opposite charge. Sound familiar? The parallels are tantalizing.

Mite

Next we apply our LCD analysis to the IVM. The procedure—calling for progressive subdivision in search of the minimum repeating unit—comes to an end with a unit that can no longer be symmetrically divided. As before, we seek the smallest system that can be duplicated to recreate the whole IVM—a microcosm, containing all the ingredients of the macro-array.

Having already split tetrahedra into equal quarters and octahedra into eighths, we skip directly to a unit consisting of a quarter tetrahedron and an eighth-octahedron back to back, sharing an equilateral-triangle face. This unit, which connects the geometrical centers (or cg's) of any adjacent tetrahedron and octahedron in the IVM, exhibits the same threefold symmetry as its two triangular pyramids taken separately. Final subdivision thus yields six equivalent asymmetrical tetrahedra: three positive and three negative (Fig. 13-4).

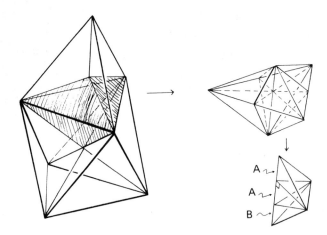

Fig. 13-4. LCD of IVM: "Mite."

As the smallest repeating unit of the IVM, this system is the minimum space filler, thus inspiring Fuller's term "Mite" (MInimum space-filling TEtrahedron.) A collection of A-modules cannot fill space; they can only make regular tetrahedra. The skinny irregular B's cannot even create a symmetrical polyhedron by themselves—let alone fill space. The Mite is the LCD of the omnisymmetrical space-filling matrix, and therefore is the minimum case, the first all-space filler:

954.09 We find the Mite tetrahedron... to be the smallest, simplest, geometrically possible (volume, field, or charge), allspace-filling module of the isotropic vector matrix of Universe.

Knowing that the Mite encompasses the asymmetrical units of both the tetrahedron and octahedron, we can identify its constituent A- and B-modules. The tetrahedron contributes an A-module, while the adjacent octahedron adds both another A (the mirror image of the first) and a B, for a total of three equivolume Modules. The positive and negative A's are in balance, and the solo B may be either positive or negative, thereby determining the sign of the whole Mite. Like A- and B-modules, Mites come in one of two possible orientations (Fig. 13-5a).

We cannot fail to comment on another 2 : 1 ratio just displayed by the minimum space-filler. As suggested by this discovery, it turns out that there must be two A-modules for every B-module in any space-filling polyhedron. We shall see how this rule applies to our familiar candidates below, and once again, Loeb's "Contribution" provides a more thorough analysis of the phenomenon. Appropriate

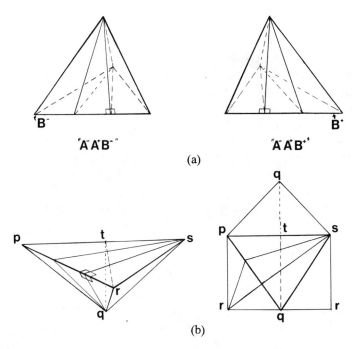

Fig. 13-5. (a) Orientations of Mite. (b) Unexpected mirror plane in Mite.

examples will be cited throughout this chapter, but readers are encouraged to turn to the back of *Synergetics* for Loeb's report.[2]

Mirrors

There's more to the Mite than meets the eye. Figure 13-5a shows a corresponding face of both a positive and a negative Mite. As every Mite incorporates both a positive and a negative A-module, it is the unpaired B-module—slanting off to one side or the other—that determines the charge of the overall system.

From the above recipe, it appears evident that this minimum space-filling tetrahedron is either right- or left-handed. This was also evident from our initial generating procedure, subdividing the IVM to carve out the smallest repeating unit with mirror symmetry, which is then split into equivalent mirror-image (positive and negative) Mites.

However, a surprising and easily overlooked result of joining two A's and a B modifies the above conclusion. An unexpected mirror plane runs through the middle of this irregular tetrahedron that we have assumed must be either right- or left-handed. This means that a

Mite is actually its own mirror image. The positive and negative versions are identical. The slanting orientation depicted in Figure 13-5a obscures this fact; however, it turns out that the Mite has two isoceles faces, a fact that indicates that the remaining two faces must be identical (but mirror-image) triangles. Therefore, the outside container of the Mite (ignoring the arrangement of its modular ingredients) incorporates a subtle mirror plane and is actually exactly the same shape as its mirror image. $A^+A^-B^+$ equals $A^+A^-B^-$ (Fig. 13-5b). This extraordinary fact means that we don't need positive *and* negative Mites to fill all space; one or the other version —or both in random combinations—will suffice. As its own mirror image, any Mite can fill either position, positive or negative.

We can conclude therefore that the two versions are identical; however, Bucky points out that different internal configurations cannot be ignored, for they point to different energy characteristics:

> Though outwardly conformed identically with one another, the Mites are always either positively or negatively biased internally with respect to their energy valving (amplifying, chocking, cutting off, and holding) proclivities... (954.43)

Cubes into Mites

Once the A- and B-game gets going, significant relationships are uncovered at every move. The game thus becomes more and more fascinating as it is played. For example, we might carve open a cube, generating six square-based pyramids, one from each face to the cg. Then slice each square pyramid into quarters, like a peanut-butter sandwich. We thereby rediscover the Mite, observing that the cube consists of twenty-four Mites (Fig. 13-6a). Oriented with one of its isosceles triangles (45°–45°–90°) on the cube's surface, the Mite simply rotates to transform Octet Trusses into boxes.

Rhombic Dodecahedra

Next, we apply the LCD procedure to the rhombic dodecahedron. Its twelve faces each frame a diamond-shaped valley ending at the cg point. Each of these inverted pyramids can be split into four (two positive, two negative) asymmetrical tetrahedra (Fig. 13-6b). This is the LCD of Fuller's "spheric," and it is none other than the Mite, the LCD of the isotropic vector matrix. Forty-eight Mites in yet another orientation make up this space-filling shape. Not surprisingly, space-filling and Mites have an important connection.

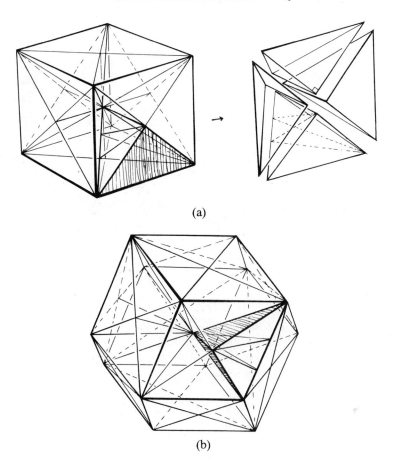

(a)

(b)

Fig. 13-6. The LCD of both cube and rhombic dodecahedron is the Mite.

No longer caught off guard by such interconnectedness, we conclude by observing that the Mite—faithfully representing octet symmetry—is also an integral component of the cube and the rhombic dodecahedron.

Coupler

Fuller's *coupler* is an irregular octahedron made of eight Mites—or sixteen A-modules and eight B-modules (Fig. 13-7). Given this composition, the "semisymmetrical" coupler is clearly a space filler. Because of the Mite's newfound mirror symmetry, the coupler does not have to have equal numbers of positive and negative B-modules; *any* eight Mites will make a coupler. This special octahedron has two

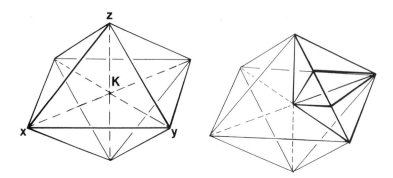

Fig. 13-7. Each Mite is one octant of the coupler.

equal-length axes (which will be referred to as x and y) and a third shorter axis (z). The point of intersection of the three axes will be called K.

The equal x and y axes outline a square equatorial cross-section, which we can now identify as the face of a cube. In fact, the coupler is two sixth-cube pyramids back to back.[3]

The other cross-sections (xz and yz) are diamonds—in fact, the exact shape of a rhombic dodecahedron's face. At this point we shall not be surprised to learn that splitting the coupler in half along either diamond cross-section isolates one-twelfth of the rhombic dodecahedron. Six half couplers make a cube; twelve half couplers (split the other way) make a rhombic dodecahedron.

"Couplers literally couple 'everything'" (954.50). Aptly named, the new octahedron joins together both pairs of cubes and pairs of rhombic dodecahedra. Fuller's nomenclature proves quite logical:

> We give it the name the Coupler because it always occurs between the adjacently matching diamond faces of all the symmetrical allspace-filling rhombic dodecahedra, the "spherics".... (954.47)

The coupler's different pairs of opposite vertices reach to the geometric centers of two adjacent polyhedra (cube or spheric, depending on the orientation), incorporating their shared face as a cross-section (Fig. 13-8a, b). Half a coupler belongs to one cube (or spheric), and the other half to its neighbor. So the coupler literally couples—well, not "everything" but—a couple of space-fillers.

Fuller continues: The coupler's role is cosmically relevant, for "rhombic dodecahedra are the unique cosmic domains of their respectively embraced unit radius closest-packed spheres" (954.47). The coupler therefore connects the centers of gravity of adjacent

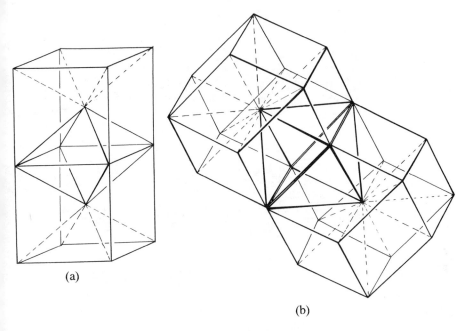

Fig. 13-8. "Coupler."

spheres, and its domain includes both the spheres and the intervening (dead air) space. This observation explains why the coupler's volumetric center was labeled K: it marks the exact "kissing point" between tangent spheres in a closepacked array. Now we have the complete story behind Fuller's somewhat dense explanation of his coupler:

> ... The uniquely asymmetrical octahedra serve most economically to join, or couple, the centers of volume of each of the 12 unit radius spheres tangentially closest packed around every closest packed sphere in Universe, with the center of volume of that omnisymmetrical, ergo nuclear, sphere. (954.48)

Perhaps the most intriguing aspect of the coupler is its similar role in the cube and in the rhombic dodecahedron, thereby linking (or coupling) the two space fillers in a new partnership.

Volume and Energy

The inventory of twenty-four modules (sixteen A, eight B) indicate that the coupler has the same volume as the regular tetrahedron with its twenty-four A-quanta. Both have a tetrahedral volume of one. (Recall also that a coupler consists of two sixth-cube pyramids back

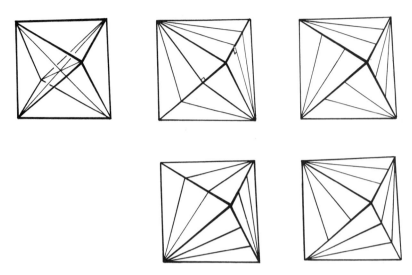

Fig. 13-9. Different rearrangements of eight Mites in the coupler.

to back, or $\frac{1}{3}$ of a cube, and that in Chapter 10 the volume of a tetrahedron was shown to be $\frac{1}{3}$ that of a cube.) Dissimilar in symmetry and shape, they are related by their shared unit volume, inviting comparison. The coupler is a different sort of minimum system: a semisymmetrical space filler, in contrast with the tetrahedron's origin as the minimum system of any kind, i.e., the first case of insideness and outsideness.

Finally, we have to comment on the coupler's internal flexibility. The number of different ways to arrange eight Mites is greatly increased by the unexpected mirror symmetry. Since positive and negative Mites can switch places, we actually have a pool of sixteen from which to choose for each of the eight positions. A coupler might consist of four positive and four negative Mites, or all positive, all negative, or any of the possible combinations in between: (0–8, 1–7, 2–6, 3–5, 4–4, ..., 8–0). Then, within each of these nine possible groups, the Mites can be switched around into four different arrangements. (A few of these combinations are shown in Figure 13-9.) The resulting 36 varieties of couplers all have the same outward shape, but in Fuller's view, their internal variations represent important distinctions in energy behavior:

When we discover the many rearrangements within the uniquely asymmetric Coupler octahedra of volume one permitted by the unique self-interorientability of the *A* and *B* Modules without any manifest of external conformation alteration, we find that under some arrangements they are abetting the *X* axis interconnectings between nuclear spheres and their 12 closest-packed... spheres, or the Y axis interconnectings.... (954.58)

When we consider that each of the eight couplers which surround each nuclear coupler may consist of any of 36 different AAB intramural orientations, we comprehend that the number of potentially unique nucleus and nuclear-shell inter-patternings is adequate to account for all chemical element isotopal variations... as well as accommodation... for all the nuclear substructurings, while doing so by omnirational quantation.... (954.54)

In other words, "energy" travels in one direction or another depending on the arrangement of the (oppositely biased) A's and B's. Synergetics thus accounts for nature's incredible variety and complexity despite its small number of different constituents. The secondary level of organization involves grouping the different couplers together and results in an explosion of potential variations. A- and B-modules can thus be rearranged into orderly octet configurations in myriad ways; clusters might appear quite chaotic in some locations and precisely ordered into whole octahedra and tetrahedra in others, while the overall space-filling matrix remains intact. As in genetics, a small number of simple constituents are able to generate a virtually unlimited repertoire of patterns.

Review: All-Space Fillers

Dismantling the cube, we saw that each of its six inverted pyramids breaks down into four Mites. What does this tell us about A- and B-modules? 24 Mites yield a total of 72 modules, with twice as many A's and B's (or 48 : 24). This two-to-one ratio gets to the modular heart of the octahedron–tetrahedron prerequisite for space filling, as developed in the last chapter.

Neither tetrahedra with only A's nor octahedra with equal numbers of A's and B's can qualify as space fillers. So far, so good. Thinking only in terms of A- and B-modules, what proportion of tetrahedra and octahedra would we need to satisfy the recipe of two A's for every B? The octahedron's 48 B's must co-occur with 96 A's; the octahedron itself supplies half of them, but 48 A-modules are still missing. Two tetrahedra will provide exactly the right number of A-modules to complete the formula, thus reconfirming the one-to-two octahedron–tetrahedron ratio discovered in the previous chapter.

The 48 Mites in the rhombic dodecahedron are easily disected into 96 A's and 48 B's, or a total of 144 modules. With twice as many A's as B's, Fuller's "spheric" does not contradict the growing evidence of a general rule.

A- and B-modules in the truncated octahedron—also a space filler —can be counted by recalling the numbers of internal single-frequency tetrahedra and octahedra determined in the last chapter. 32 tetrahedra, with 24 A-modules each, contribute 768 A's, while 16 octahedra consist of 768 A's and 768 B's, for a total of 1536 A's and 768 B's altogether (768 × 2 = 1536). We can begin to have confidence in the reliability of our 2 : 1 ratio, especially in view of the jump to much larger numbers. With a total of 2304 modules, this is the smallest truncated octahedron outlined by IVM vertices. Fuller's inventory of all-space fillers (954.10) lists the truncated octahedron with twice the linear dimensions and eight times the volume (consisting of 18,432 quanta modules) rather than the smaller version. The reason for this choice is not clear; however, the magnitude of these quantities hints at the complexity of his modular system—in terms of both the variety of systems that can be made from the modules and the number of rearrangements within those systems. From these numbers it is evident that, although the quanta modules help us to understand the conceptual essence of many polyhedral intertransformabilities, they are very impractical for hands-on experimentation. So many modules are needed to make complete polyhedra that this system does not offer an ideal strategy for model making. Instead, we might utilize the analysis to work out relationships on paper.

From the above examples, we can see that the A- and B-modules get to the root of the two-tetrahedron–one-octahedron rule developed in the previous chapter. Because the earlier analysis depended on disassembling its two basic units, it was necessary to probe further to isolate the real quanta. A- and B-modules, which cannot be symmetrically subdivided, were isolated as the true quanta with which to measure and analyze related polyhedral systems.

Impressed by the geometric significance of these modules, Fuller proposes that somewhere within this discovery lie secrets with far greater applicability than just to geometry:

The *A* and *B* Quanta Modules may possibly quantize our total experience. It is a phenomenal matter to discover asymmetrical polyhedral units of geometry that are reorientably compositable to occupy one asymmetrical polyhedral space; it is equally unique that, despite disparate asymmetric polyhedral form, both have the same volume.... Their unit volume and energy quanta values provide a geometry elucidating both fundamental structuring and fundamental and complex intertransformings, both gravitational and radiational. (920.01)

From their energy associations to their remarkable symmetry, these modules synthesize much of Fuller's research. Significant relationships to physical phenomena may well reward continued investigation, for nature also deals with discrete quanta, creating endless variation through synergetic recombinations. Fuller reasoned that his geometric quanta—the end result of a systematic and logical progression of steps—must relate to physical phenomena. The approach is typically Fuller's: assume significance until proven otherwise. In essence he suggests that tiny whole or discrete systems should replace irrational unending digits—somehow providing a comprehensive rational coordinate system.

Cosmic Railroad Tracks: Great Circles

Any planar closed loop drawn on the surface of a sphere is necessarily a perfect circle, as a result of the sphere's steady curvature. Such loops qualify as either "great" or "lesser" circles, and the distinction is defined in mathematics as follows. A *great circle* is formed by the intersection of a plane passing through the center of a sphere with the surface of that sphere. Any other circle, no matter what size, is *lesser*. The center of a great circle coincides with the sphere's center. In short, a great circle is an equator—found in any angular orientation, but always around the fattest part of its sphere.

What's so great about a great circle? Above all, it provides the shortest route between any two points on a sphere. This geometric fact is not obvious in many cases; for example, looking at a globe it appears that the logical route between two points situated some distance apart on the Tropic of Cancer involves traveling along their shared "lesser circle" band (Fig. 14-1a). However, the principle is made more obvious by Fuller's juxtaposition of two extreme cases. He describes a small lesser circle near the North Pole of an imaginary globe, and labels two points *A* and *B* (Fig. 14-1b). As with the larger Tropic of Cancer, the eye naturally travels from *A* to *B* along their mutual lesser-circle path, without suspecting that this represents the "long way around." Fuller then redraws the same lesser circle in a new location—superimposed over the globe's equator so that *A* and *B* both fall on the horizontal great circle. The shortest route between *A* and *B* is suddenly obvious. Likewise, between *any* two points on a sphere the most expedient route will be a great-circle segment; Fuller's example makes it easy to see that a lesser-circle path will always present a detour.

Other special characteristics: Any two great circles on a sphere must intersect twice—specifically, at two points 180 degrees apart. There is no other possibility: unless they are actually the same circle, two great circles can neither avoid each other altogether, nor intersect only once, nor intersect more than twice. Finally, the junction of two great circles inscribes two pairs of equal and opposite angles on

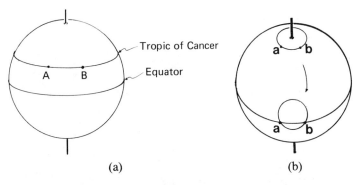

(a)   (b)

Fig. 14-1

the sphere's surface, whose two angular values add up to 180 degrees; the statement is equivalent for the intersection of two Euclidean "straight lines." Lesser circles do not share this property; in fact, their intersection produces opposite angles which are necessarily unequal. In conclusion, on the surface of a sphere, only great circles have the geometric characteristics of "straight lines."

As we examine different aspects of great circles, you will notice that much of this material is quite complicated; the patterns are too intricate to be readily visualized in the mind's eye. Furthermore, connections drawn between the geometric models and physical phenomena are unusually speculative. However, as Fuller's *Synergetics* devotes considerable attention to great circles, anyone who has tried to decipher these sections will welcome full coverage. Referring back and forth from the text to the drawings will be essential.

Why Are We Talking About Spheres?

The vertices of regular and semiregular polyhedra lie on the surface of an imaginary sphere, which is to say that all vertices are equidistant from a polyhedron's center. Given this fact, we can picture spherical versions of each polyhedron, in which the polyhedral edges have stretched outward to become great-circle arcs and the faces have expanded into curved surfaces, as if each shape had been blown up like a balloon. Figure 14-2 shows a spherical tetrahedron, octahedron, and icosahedron as examples. Comparing these systems with their planar counterparts, it is clear that polyhedral edges are actually chords of great-circle arcs. We can conclude that the shortest distance between two events of a system always involves a great circle.

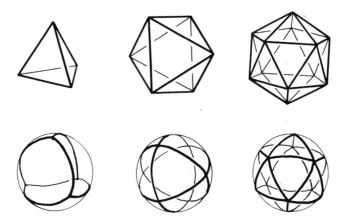

Fig. 14-2. Spherical polyhedra.

The concept (rather than the reality) of a sphere—i.e., an omnisymmetrical container—acts as a frame of reference for polyhedral systems. Spherical polyhedra thus introduce new versions of familiar characters. The topological information (that is, the numbers and valencies of vertices, edges, and faces) of any polyhedron are displayed on a spherical canvas. An obvious consequence of this type of display is that shape is no longer a variable. Shape similarities, which are so rigorously accounted for by A and B modules, are thus ignored; our investigation now focuses in on topological, or surface, characteristics. Transforming polyhedra into balloons temporarily equalizes shape and size, providing a "common denominator" for other comparisons. The process develops a somewhat unorthodox chart.

New Classification System

However, the chart is not yet complete. Simply projecting edges and faces onto a spherical surface does not teach us anything new. We have yet to exploit the nature of the sphere.

Spheres suggest spin. That's how synergetics initially arrives at the omnidirectional form. Spin any system in all directions, and ultimately the action will have defined a circumscribing spherical envelope. Fuller places considerable emphasis on the "spinnability" of systems, arguing that as everything in Universe is in motion, the different axes of spin inherent in systems are worthy of investigation.

All polyhedra have three sets of topological aspects: vertices, edges, and faces. These sets correspond to three types of axes of rotational symmetry (or "spin") which connect pairs of either polar-opposite vertices, mid-edge points, or face centers. As a polyhedron spins about any one of these axes, an implied great circle is generated midway between the two poles, in other words, at the equator. Equators corresponding to all existing axes of symmetry can be simultaneously represented on a spherical surface, creating an exhaustive chart of the topological symmetries of a given system. Each symmetrical polyhedron has its own great-circle diagram, which incorporates all its axes of rotational symmetry—or axes of spin, in Fuller's terminology. The patterns generated by related polyhedra may include some of the same circles, as determined by common symmetries; however, the complete chart of a polyhedron is exactly shared only by its dual, as will be shown below.

Great circles reveal a new aspect of polyhedral "intertransformability," a novel (if obscure) means of detecting symmetrical relationships among systems. We can anticipate the emergence of decidedly unfamiliar patterns in this game for their resemblance to the source polyhedron is sometimes subtle. Finally, in pursuing this study we discover new variations on upper limits and minimum cases. Great circles thus provide another tool with which to detect inherent spatial constraints.

Great-Circle Patterns

Let's begin with a representative system, the octahedron. We interconnect polar opposites, starting with vertices, followed by mid-edge points and finally face centers. Six paired vertices are connected by three mutually perpendicular lines, the familiar XYZ axes meeting at the octahedral center of gravity (Fig. 14-3a). These axes define three orthogonal great circles, which divide the sphere's surface into eight triangular areas, or octants. Three symmetrically arranged great circles will always form the edges of a spherical octahedron (Fig. 14-2, middle). That much is straightforward.

On to edges. The octahedron's twelve edges consist of six opposing pairs, the midpoints of which can be connected by six intersecting axes (Fig. 14-3b). The same number of great circles are thereby generated, delineating another facet of the octahedron's symmetry (Fig. 14-4). Unlike the previous case, the pattern made by six great circles does not look like an octahedron. Its twenty-four right

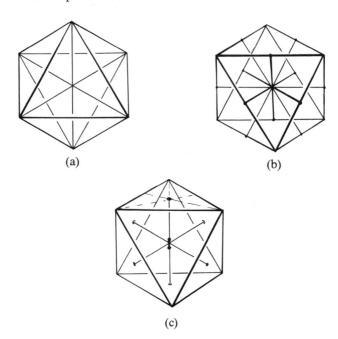

(a)

(b)

(c)

Fig. 14-3. Different sets of axes of symmetry. (a) Paired vertices yield 3 axes of four-fold rotational symmetry. (b) Paired edges yield 6 two-fold axes. (c) Paired faces yield 4 three-fold axes.

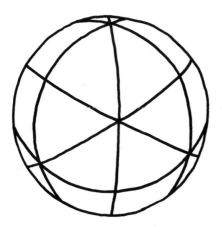

Fig. 14-4. Six great circles of the octahedron.

isosceles triangles[1] outline the edges of both the spherical cube and rhombic dodecahedron, as well as the edges of two intersecting spherical tetrahedra (otherwise known as the "star tetrahedron"), thus highlighting the topological relationship between these four systems. This exercise demonstrates a new aspect of "intertransformability": great circles generated by a given polyhedron often delineate the spherical edges of its symmetrical cousins.

Fig. 14-5. Four great circles.

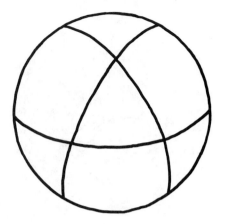

Finally, the centers of opposite faces are joined together (Fig. 14-3c). The octahedron's four pairs of triangles define four intersecting axes, and in turn four symmetrically arrayed great circles (Fig. 14-5). As the spherical edges of Fuller's vector equilibrium, this pattern is particularly significant, as will be developed below.

The sets of three, six, and four great circles can all be superimposed on one sphere to exhaustively display the "unique topological aspects" of the octahedron. With a grand total of thirteen, we have located all the great circles that correspond to the octahedron's symmetry. We can go no further.

Now consider the cube. It is quickly apparent that its total great-circle pattern will be identical to that of the octahedron. The cube's eight vertices generate the same four great circles as the octahedron's faces, its six faces correspond to the three orthogonal great circles, and the twelve edge midpoints are identically situated to those of the octahedron. Such is the result of duality. The same sets of circles are generated by different elements, and the end results are equivalent.

Next, we turn to the VE. With two kinds of faces the situation might seem more complicated; however, the above procedure still applies. The VE's eight triangles define the same four axes as the faces of the octahedron, while its squares contribute three orthogonal (XYZ) axes. Seven great circles altogether are generated by the axes of the fourteen VE faces. Twelve vertices correspond to the same six great circles as those of the octahedron (or cube) edges, and finally 24 edges spin out the unfamiliar pattern of twelve great circles. With a total of 25 great circles, the topological parameters of the VE are exhausted (Fig. 14-6).

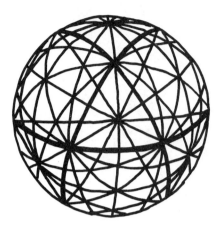

Fig. 14-6. Twenty-five great circles of the VE.

The tetrahedron presents a slightly different situation in that its vertices do not group into polar opposites, but rather are positioned directly across from the centers of faces. However, four axes of symmetry (all going through the center of gravity) can be created by connecting the vertices with their opposite faces. In having axes of rotational symmetry that connect faces and vertices, the tetrahedron is again unique; as we recall from Chapter 4, only tetrahedra have the same number of vertices as faces. (Only the tetrahedron is its own dual.) The four great circles generated by these unorthodox axes produce (once again) the spherical vector equilibrium. This is not surprising if we recall the lesson from "multiplication by division," which first uncovered this shared symmetry between the tetrahedron, octahedron, and VE: By simply interconnecting mid-edge points, all three polyhedra were found to be inherent in the topological makeup of the starting-point tetrahedron. They share the same four axes of symmetry.

The axes of symmetry associated with the tetrahedron's six edges are more orthodox: mid-edge points of opposite edges are simply joined to reestablish the *XYZ* axes. Familiar by now with evidence of right angles hiding within this triangular shape, we can no longer be caught off guard by this discovery. The corresponding three great circles are the edges of the spherical octahedron, once again illustrating the depth of the octahedron–tetrahedron relationship. And now all topological aspects are used up; the tetrahedron has seven great circles, the minimum number possible for symmetrical polyhedra.

Next, we look at the maximum case. It's not easy to accept the concept of an upper limit on the number of symmetrically positioned great circles that can be imposed on a sphere; common sense suggests that we should be able to keep adding new rings inde-

Fig. 14-7. Six great circles of the icosahedron.

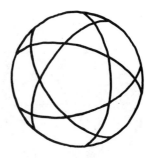

finitely. However, we recall from Chapter 3 that all systems are polyhedral—that is, everything that divides inside from outside can be described in terms of some number (four or more) of "event complexes" and their relationships—and from Chapter 4 that the system with the greatest number of identical regular polygons and equivalent vertices is an icosahedron. This tells us that the number of great circles allowed by the topological aspects of the icosahedron is the maximum for these symmetrical patterns.

The axes defined by the icosahedron's twelve vertices introduce six great circles. We already have a pattern with six circles (octahedron–VE, Fig. 14-4); however, this is an entirely new set, "out of phase" with the earlier group, as defined by the "jitterbug" transformation (Fig. 14-7). Outlining 12 pentagons and 20 triangles (as opposed to 24 isosceles triangles), these arcs present the most symmetrical arrangement of six great circles. When a vector equilibrium contracts into an icosahedron in the jitterbug transformation, the radius–edge-length equivalence is lost, but the distances between adjacent vertices on the surface are suddenly all equal. As an icosahedron produces the most symmetrical distribution of twelve vertices on a closed system, it follows that the same is true for their corresponding six great circles.

Next, the axes of symmetry connecting the centers of icosahedron faces generate ten great circles, while the thirty edges spin out fifteen more: $6 + 10 + 15 = 31$, the total number of great circles in the limit-case pattern (Fig. 14-8). These circles provide the spherical edges of a pentagonal dodecahedron as well as those of an icosahedron, and (less predictably) also include the octahedron. These relationships will be explored below.

Least Common Denominator

The spherical icosahedron divides the surface of a sphere into the greatest number of completely regular domains (with both equal arc lengths and equal surface angles). We can therefore subdivide one of

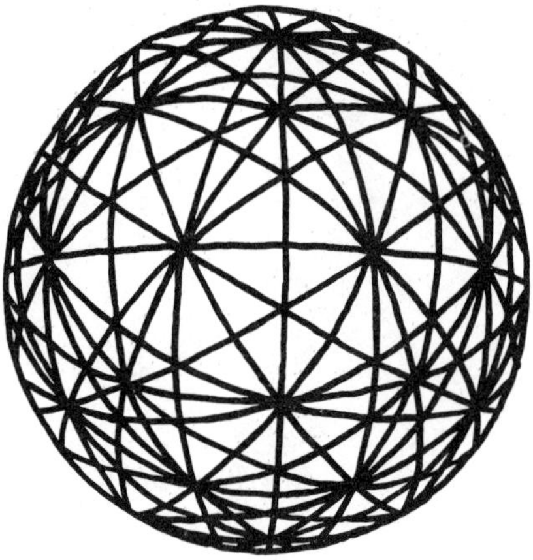

Fig. 14-8. Thirty-one great circles.

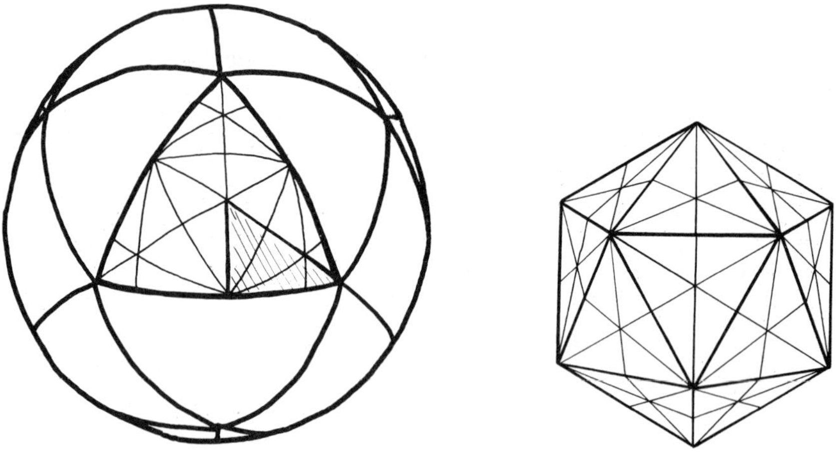

(a)   (b)

Fig. 14-9. Maximum of 120 equivalent domains: LCD.

these symmetrical triangles to find the least common denominator of surface unity—in much the same way we isolated A and B modules. Each triangle is split into six equal parts by perpendicular bisectors, to obtain a final non-symmetrically-divisible unit (Fig. 14-9a). The result is 120 asymmetrical triangles (60 positive, 60 negative), the maximum number of equivalent domains on the surface of a sphere. These perpendicular bisectors are the icosahedron's fifteen great circles (Fig. 14-9b).

Imagine that some number of hypothetical creatures are to inhabit the surface of a large sphere and it is considered necessary that each one sit in the middle of an identical plot. The consequence of this stipulation is that no more than 120 creatures can fit on the sphere, no matter how large it is. This result is certainly counterintuitive, for assuming a large enough sphere, it seems that we should be able to accommodate as many creatures as we want. However, the unyielding laws of symmetry limit the population to the unexpectedly low number of 120.

We go back to the 31-great-circle diagram to observe the 120 triangles in context. These asymmetrical triangles are true LCD units; any one of them contains all the geometric information necessary to reconstruct the entire pattern. This is an important factor in the development of geodesic domes, as will be seen in the next chapter, which discusses the relationship of great circles to geodesic domes.

LCD: "Intertransformability"

The following exploration is similar to the transmutations of A and B modules and uncovers some of the same relationships; however, consistent with great-circle limitations, this study deals only with surface characteristics. Each symmetrical great-circle pattern has a least common denominator. For example, the spherical octahedron has eight equilateral faces, which can be split into six asymmetrical triangles, each one $\frac{1}{6}$ of $\frac{1}{8}$, or $\frac{1}{48}$, of the whole surface. These triangles are LCD units, because they cannot be further subdivided to yield equivalent shapes.

To isolate the LCD of 25 great circles, we must take into consideration that the spherical VE has two types of faces. Therefore, the smallest unit that can be reproduced to generate the whole pattern

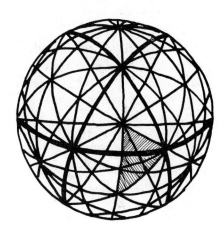

Fig. 14-10. LCD unit.

requires $\frac{1}{6}$ of a VE triangle joined to $\frac{1}{8}$ of an adjacent square. In this way, both aspects of the VE pattern are incorporated into the LCD, and the result is an asymmetrical triangle that covers $\frac{1}{48}$ of the sphere (Fig. 14-10).

By changing the boundaries, LCDs of a given spherical polyhedron often can be caused to make up the faces of various other polyhedra. Interesting transformations are found between the great circles of the VE and the icosahedron, for the shift from 25 to 31 is another result of the icosahedron's "out of phase" role in the "cosmic hierarchy."

LCD of 31 Great Circles

Each triangle of the icosahedron consists of six LCDs, for a total of 120 asymmetrical triangles. That much is straightforward, as are the first two transformations.

Instead of the standard groups of six LCDs making up icosahedron faces, we change the boundaries. Four of these units form the diamond face of the spherical *rhombic triacontahedron*, so that we have thirty groups of four triangles instead of twenty groups of six. Figure 14-11a shows that each icosahedron edge is the long diagonal of one of the thirty diamonds; the planar rhombic triacontahedron is thus a "degenerately stellated" icosahedron.

Secondly, we can recombine the 120 units into twelve groups of ten with each triangular unit radiating out from an icosahedron vertex, to highlight the pentagonal dodecahedron (Fig. 14-11b).

Next, we discover a few asymmetrical transformations. A spherical triangle of fifteen LCD units, incorporating a complete icosahedron triangle radially framed (like a pinwheel) by nine extra LCDs, is one face of the spherical octahedron. With this observation, we isolate the octahedron face just by looking at the pattern, so it's worth checking the arithmetic: multiply fifteen units per face by eight faces to get 120, or the whole system. Figure 14-11c shows this skew relationship, again reminiscent of the jitterbug transformation. And lastly, we observe the edges of a spherical VE. At first this seems to present a contradiction, given the unsynchronized relationship of the 25- and 31-great-circle patterns. However, the VE's four great circles are included among the 31 icosahedral equators, although asymmetrically positioned with respect to its vertices (Fig. 14-11d). We now recall from Chapter 11 that it was possible for all the faces of an octahedron to be aligned with eight of the icosahedron's twenty (Fig. 11-5). This skew correspondance, which defined the S module, shows

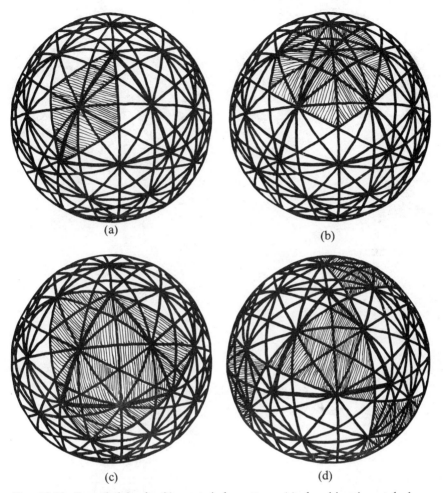

Fig. 14-11. Revealed in the 31-great-circle pattern: (a) rhombic triacontahedron; (b) pentagonal dodecahedron; (c) octahedron; and (d) VE.

how a subset of four out of the ten circles generated by icosahedron faces will be correctly situated to create the spherical VE.

The number of different polyhedra hiding within the 31 great circles reemphasizes the existence of significant relationships between the "out of phase" icosahedral family and the IVM group.

VE's 25 Great Circles

The intertransformability displayed by the VE's least common denominator is straightforward, in contrast to the skew relationships demonstrated above. Groups of four units create diamond faces

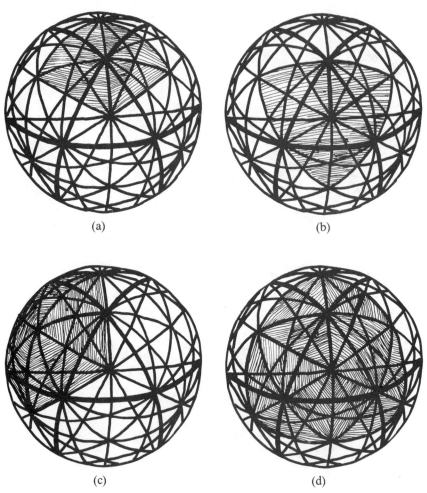

(a)

(b)

(c)

(d)

Fig. 14-12. Revealed in 25 great-circle pattern: (a) rhombic dodecahedron; (b) octahedron; (c) cube; and (d) tetrahedron.

exactly centered over each VE vertex, thereby defining the twelve faces of its dual, the rhombic dodecahedron (Fig. 14-12a). Six LCD triangles come together to create octahedron faces (Fig. 14-12b), while the cube's six squares each consists of eight asymmetrical units (Fig. 14-12c). And a spherical tetrahedron uses a dozen LCDs per face, distributing the 48 units among only four faces (Fig. 14-12d).

This brief look at various regroupings of LCD units shows how great-circle diagrams provide a new way to classify certain polyhedral characteristics, and thereby discover shared symmetries between systems.

Operational Mathematics

Fuller made a remarkable discovery about great-circle patterns that is responsible for their great significance in his mathematics. This discovery involves an intricate relationship between central and surface angles and could so easily be missed that one cannot help reflecting on the intuition that led Fuller to such an insight. As with other aspects of his "operational" method, the demonstration relies on readily available materials, but its significance extends to structuring in nature. This is a particularly satisfying exercise, and readers are encouraged to make Bucky's discovery themselves by following the simple instructions. Rather than a "plane," Bucky starts with a real system, a "finite piece of paper" (831.01). Using a compass, draw four circles with diameter of approximately 6 inches, and then cut them out with scissors. Fold each one. Then fold the resulting semicircles into thirds, as shown in Figure 14-13. The section labeled *A* is folded toward you, while *C* is folded back, producing a Z-shaped cross-section.

Making sure that all creases are sharp, unfold the paper to obtain a circle with three intersecting diameters clearly marked by fold lines (Fig. 14-13). The circle is thus divided into six equilateral triangles. The process of sweeping out a circle with a compass insures that all radii are equal, and because at one point in the procedure all six pie-slices are piled up together, we know that the central angles must

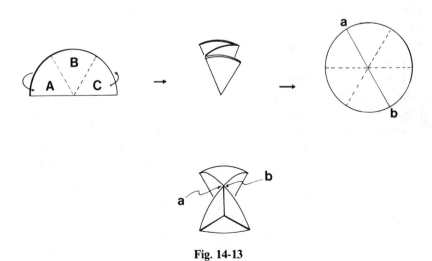

Fig. 14-13

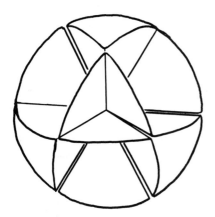

Fig. 14-14. Four bowties create the four great circles of the VE.

all be exactly the same. 360 degrees divided among six equal segments yields 60-degree angles. First-hand experience has confirmed both the constant radius and sixty-degree angles, and therefore the presence of equilateral triangles with arc segments at their outer edges is experientially proved.

One fold (line *AB*) faces you; the other two folds are facing away. Bringing point *A* to point *B*, we create one of Fuller's "bowties" (Fig. 14-13, bottom). A bobby pin straddling the seam keeps the bowtie together: two unit-length regular tetrahedra joined by an edge. We repeat the procedure three times, producing four bowties in all. It is then apparent that two of them can be placed seam to seam and pinned together with two more bobby pins, to produce four tetrahedra surrounding a half-octahedral cavity. The other two bowties are similarly paired, and finally the two pairs are connected along their congruent fold lines with four more bobby pins. A complete paper sphere emerges (Fig. 14-14). This strange procedure has created a very familiar pattern: four continuous great circles, or a spherical VE. The only materials required are four paper circles and twelve bobby pins.

What has happened? Four separate paper circles have been folded, bent, transformed into bowties, and pinned together without paying any attention to converging angles. No special jig is required to line up adjacent bowties and insure that consecutive great-circle arcs are collinear. Folded edges are simply brought together, and four continuous great circles magically reappear, as if the original paper circles were still intact. Looking only at the finished model, it appears that we had to cleverly cut slits in the paper circles in just the right places to allow the four circles to pass through each other. The procedure is reminiscent of the magic trick in which a hankerchief is cut into many tiny pieces and thrown randomly into a hat, only to reappear intact.

A spherical VE can be constructed through this simple folding exercise because of the specific interplay of its surface and central angles. Remarkable numerical cooperation is required to allow adjacent central angles to fold out of flat circular disks, while automatically generating correct surface angles. That four great circles will submit to this bowtie operation is not at all obvious from studying the whole pattern, and even less so in the case of other, far more complicated models.

Conservation of Angle

Physics tells us that a beam of light directed toward a mirror at some angle from the left will bounce away from the mirror making the same angle on the right. The angle of incidence is equal to the angle of reflection. Fuller points out that the same is true for the great-circle models, if we think of the paths as trajectories.

Great circles maintain the illusion of being continuous bands, argues Fuller; however the bowtie procedure reveals the truth about these patterns, and in so doing illustrates an aspect of energy-event reality. A great-circle path may look continuous, but what really happens is that as soon as a trajectory (or great-circle arc) meets an obstacle, in the form of another great-circle event, they collide and the course of both trajectories is necessarily altered. Both paths are forced to bounce back, just like a ball bouncing off a wall.

Here's the fun part. Because the intersection of two great circles provide two pairs of equal angles, their paths mimic the classical collision in physics. The same angular situation results from two overlapping great circles as from an idealized "energy-event" collision; these symmetrical patterns can therefore be thought of as the paths taken by billiard balls on the surface of a spherical pool table. If you didn't see the collisions, you might think that two balls went through the same point at the same time; however, their true paths are bowties.

Imagine a great-circle wall constructed vertically out from the surface of a sphere. If a ball traveling parallel to the sphere's surface (describing another great circle) hit the wall and bounced back, "...it would bounce inwardly off that wall at the same angle that it would have traversed the great-circle line had the wall not been there..." (901.13; Fig. 14-15a). Adjacent bow-ties therefore produce collinear great-circle arcs, because the angle made by the paper disk

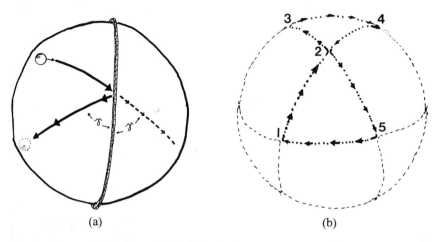

(a)   (b)

Fig. 14-15. (a) Ball bounces back from the wall at the same angle it would have made with the wall on the other side had the wall not been there. (b) "Local holding pattern:" figure-eight loop.

"bouncing back" from a bobby-pin collision point is the same angle the circle would have made on the other side if the bobby pin had not been there. Fuller continues, "and it would bounce angularly off successively encountered walls in a similar-triangle manner..." (901.13). Hence the completed bowties. The image of the colliding ball and great-circle wall makes it easier to understand why Fuller interpreted great circles as the trajectories of energy events, and explains why he was convinced that these models have physical significance. His bowtie analogy is consistent with the classical collision model.

Consider the paper model described above. Soon after an energy event meets its first collision, the new trajectory meets a second obstacle and its course is again altered. Conservation of angle determines the new heading once again, and the process repeats until, at the sixth collision, the great-circle path comes back upon itself, completing the bowtie loop (Fig. 14-15b).

Successive arc segments of one energy event form figure-eight loops, or "local holding patterns" (455.05), which lie side by side and appear to be continuous great circles. Construction paper shows us how it works, and Fuller tells us that this is what happens in physical reality as well. Discrete energy events form local circuits, just as discrete paper circles form bowties; they only *look* continuous.

Foldable Systems

Just as we needed at least four vertices to make a system, "four is also the minimum number of great circles that may be folded into local bow ties and fastened corner-to-corner to make the whole sphere..." (455.04). An interesting characteristic of this minimum model is that the sum of the areas of its four separate circles, which fold up to create the model, is equal to the surface area of the sphere they define, or $4\pi R^2$. A system with less than four great circles cannot be directly constructed out of that number of bowties, but in some cases the pattern can be simulated using more than the prescribed number of paper disks and doubling the polyhedral edges:

> You cannot make a spherical octahedron or a spherical tetrahedron by itself...
> 109°28′ of angle cannot be broken up into 360-degree-totalling spherical
> increments.... (842.02–3)

A spherical tetrahedron can only be created out of "foldable" paper circles by constructing the six-great-circle spherical cube, which produces two intersecting tetrahedra. See Fig. 14-16: six great circles will submit to a bowtie construction, similar to the four-great-circle model described above:

> 842.04 Nor can we project the spherical octahedron by folding three great circles.
> The only way... is by making six great circles with all the edges double....

In this construction, six circles are folded in half and then the resulting semicircles are each folded in the middle at a right angle. The six bent semicircle pancakes (rather than three open bowties) are then simply pinned together to simulate the three great circles of the octahedron (Fig. 14-17).

Fuller attributes these discoveries to a "basic cosmic sixness":

> There is a basic cosmic sixness of the two sets of tetrahedra in the vector
> equilibrium. There is a basic cosmic sixness also in an octahedron minimally-great-
> circle-produced of six great circles; you can see only three because they are doubled
> up. And there are also six great circles occurring in the icosahedron. All these are
> foldable.... This sixness corresponds to our six quanta: our six vectors that make
> one quantum. (842.05–6)

Both six-great-circle patterns can be constructed out of foldable circles. The star-tetrahedron version involves six bowties with two right isosceles triangles on the surface (60°–60°–90°) created by central angles of 70°32′ (a) and 54°44′ (b), as labeled in Figure

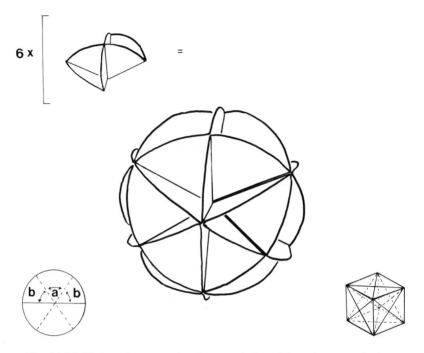

Fig. 14-16. Six bowties create the six great circles of the cube or octahedron.

14-16. The six bowties are pinned together at the seams to create the 24 surface triangles of the spherical cube and star tetrahedron. And it works: the illusion of six continuous circles is maintained.

The icosahedral six great circles fold into pentagonal bowties. Fold lines divide each circle into ten equal slices of pie, carving out the 36-degree central angles. Each circle is then pinched together at one point to form a double-pentagon figure-eight. Just like the triangular predecessors, each circle folds into a "local circuit" and is connected to other local circuits to create the illusion of continuous great circles (Fig. 14-18).

Fuller experiments with the "foldability" of considerably more intricate patterns. Larger numbers of great circles intersect more frequently, and arc segments have correspondingly smaller central angles. The number of folds and the precision required for each of the tiny irregular angles make these higher-frequency models extraordinarily difficult to build, and even more difficult to visualize without a model. Each of the ten great circles of the icosahedron can fold into winding chains of six narrow tetrahedra, which then interlink to reproduce the ten-great-circle pattern.[2] Fuller claims that

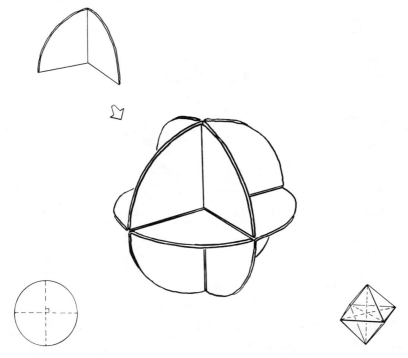

Fig. 14-17. Three-great-circle model requires six foldable circles.

Fig. 14-18. Six pentagonal bowties create the six great circles of the icosahedron.

the fifteen-great-circle pattern (outlining the 120 LCD units) can be reconstructed with fifteen four-tetrahedron chains.[3] While a paper circle will fold into four consecutive LCD tetrahedra, it is not clear how they fit together to recreate the whole sphere. We note however that this pattern can be easily generated by using thirty paper circles and doubling the edges. Each circle folds into four adjacent LCD tetrahedra, to form one self-contained diamond—a face of the rhombic triacontahedron (Fig. 14-19).

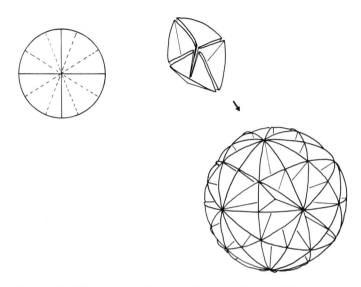

Fig. 14-19. Fifteen-great-circle model, created by doubling the edges.

For the purposes of this chapter, we want to understand the basics of how foldable great circles work; further experimentation with construction paper is left to curious (and ambitious) readers. We proceed to look at Fuller's interpretation of the significance of this behavior in terms of physical reality.

Energy Paths

Here's the basic premise: First, the concept of a sphere provides a model of the generalized system, and on the surface of a sphere the shortest route is a geodesic path. Secondly, Universe breaks down into discrete systems, which consist exclusively of energy events and their relationships. And finally, energy is always in motion, perpetually transferred between finite local systems along most direct routes, and therefore, energy must be traveling through Universe via great-circle paths.

Gas Molecules

We consider a specific example:

A vast number of molecules of gas interacting in great circles inside of a sphere will produce a number of great-circle triangles. The velocity of their accomplishment of this structural system of total intertriangulation averaging will seem to be instantaneous to the human observer. (703.14)

Fuller's hypothetical molecules are bouncing around so fast that the overall pattern of their activity seems instantaneous. But in reality, he reminds us, there is no "instant Universe"; time is always a factor. Molecules collide and bounce back; equal angles created by the symmetry of their collisions create the illusion of full great circles interweaving, but their trajectories are really local loops:

> The triangles, being dynamically resilient, mutably intertransform one another, imposing an averaging of the random-force vectors of the entire system, resulting in angular self-interstabilizing as a pattern of omnispherical symmetry. (703.14)

To Fuller the ability to fold individual paper circles and automatically generate an entire symmetrical pattern gives great-circle models important physical relevance. They demonstrate his statement that no two lines can go through the same point at the same time and illustrate the concept of "interference patterns." The model, says Fuller, is consistent with physical behavior.

Although physical systems are always imperfect, the result of a vast number of interactions is approximate symmetry. With enough data or time, all possible paths can be tried, and the properties of space come into play. What does a maximally symmetrical distribution on a spherical surface look like?

> The aggregate of all the inter-great-circlings resolve themselves typically into a regular pattern of 12 pentagons and 20 triangles, or sometimes more complexedly, into 12 pentagons, 30 hexagons, and 80 triangles described by 240 great-circle chords. (703.14)

The pattern is always icosahedral, some version of the maximally symmetrical shell. (Refer to Chapter 11 for a comparison of space-filling versus shell symmetries.) *Icosahedral symmetry fits the most great circles on a closed system.* One notable characteristic of this pattern, no matter how high the frequency of subdivision,[4] is the presence of exactly twelve pentagons evenly distributed around the system—one at the location of each five-valent icosahedral vertex: "... the 12 pentagons, and only 12, will persist as constants; also the number of triangles will occur in multiples of 20" (703.15). This constant number of pentagons, together with "twelve spheres around one" and "twelve degrees of freedom," suggests a fundamental twelveness inherent in space. We shall return to this pattern in the next chapter, looking at geodesic structures.

Fuller uses molecules bouncing around in a spherical system as a model, because molecules are energy events occurring in large enough numbers to describe the symmetrical patterns developed as polyhedral abstractions. In conclusion, energetic behavior is subject to symmetrical constraints, and in order to adhere to logic or theory, probability calls for very large numbers.

Great-Circle Railroad Tracks of Energy

Chapter 8 examined the closest packing of spheres. The next question in Fuller's investigation is how "energy" will navigate through these clusters. The idea that closepacked spheres present a sort of conceptual model of physical Universe is at the root of Fuller's great-circle studies. As a space-filling array of discrete systems, whose omnisymmetrical qualities recommend them as a general representation of eternally spinning energy-event systems, close-packed spheres do provide a tantalizing model. Chapter 8 described the closest packing, which places every sphere in contact with twelve others, and this chapter mapped out the complete network of short-est-distance paths around a spherical system described by these twelve contact points. The "cosmic railroad tracks" thus described were the 25 great circles of the VE:

> The 12 points of tangency of unit-radius spheres in closest packing, such as is employed by any given chemical element, are important because energies traveling over the surface of spheres must follow the most economical spherical surface routes, which are inherently great circle routes, and in order to travel over a series of spheres, they could pass from one sphere to another only at the 12 points of tangency of any one sphere with its closest-packed neighboring uniform-radius sphere." (452.01)

In an effort to describe a symbolic model of atomic and molecular activity, Fuller allows the omnisymmetrical (or perhaps "omni-spin-nable" is more appropriate) form of spheres to represent atoms and explains that energy, or charge, can only travel from system to system through points of tangency. (Fuller is also quick to point out that tangency is actually "extremely close proximity", for in physical reality nothing "touches.") Energy is thus found either in finite local circuits (bowties) on one system, or jumping over to a neighboring system through VE vertices:

> These four great-circle sets of the vector equilibrium [i.e. sets of 3, 4, 6, and 12] demonstrate all the shortest, most economical railroad "routes" between all the points in Universe, traveling either convexly or concavely. The physical-energy travel patterns can either follow the great-circle routes from sphere to sphere or go around in local holding patterns of figure eights on one sphere. Either is permitted and accommodated." (455.05)

These "universal railroad tracks" are specifically along the 25 great circles because of the relationship of vector equilibrium to the cosmically significant closepacked spheres.

Icosahedron as Local Shunting Circuit

"The vector equilibrium railroad tracks are trans-Universe, but the icosahedron is a locally operative system" (458.12). This distinction, introduced by the jitterbug transformation, has particular significance in Fuller's great-circle theories. The VE is integral to our space-filling network of equivalent vectors; however, once the VE contracts into the icosahedron with its slightly shorter radius, it is disconnected from that universal IVM network. It loses its contacts. "Energy" is free to travel endlessly throughout the railroad tracks of Universe, sliding from sphere to sphere along the economic great-circle paths, until it runs into an icosahedron. The icosahedral 31 great circles are not "trans-Universe" lines of supply; their function is to disconnect energy from the closest-packing railroad tracks and direct it into local orbits. The icosahedron throws the switch:

> The icosahedron's function in Universe may be to throw the switch of cosmic energy into a local shunting circuit. In the icosahedron energy gets itself locked up even more by the six great circles—which may explain why electrons are borrowable and independent of the proton-neutron group. 458.11

Fuller suggests that there might be a meaningful connection between the icosahedron and the electron, because the tiny negative charge is readily transferred from atom to atom in molecules and crystals. The icosahedron's independent role, in Fuller's view, "shunting" energy into local circuits (that is, able to disconnect energy charges from a bigger matrix) is suggestive of the electron's role:

> 458.05 The energy charge of the electron is easy to discharge from the surfaces of systems. Our 25 great circles could lock up a whole lot of energy to be discharged. The spark could jump over at this point.... If we assume that the vertexes are points of discharge, then we see how the six great circles of the icosahedron—which never get near its own vertexes—may represent the way the residual charge will always remain bold on the surface of the icosahedron.

Lacking contact points, the icosahedron is a free-floating unit in Universe; so is the electron. The suggestion of a relationship remains just that, a tantalizing parallel, seeds perhaps of future investigation, but certainly among Fuller's more abstruse parallels.

Inventory: Seven Unique Cosmic Axes of Symmetry

The VE's 25 great circles incorporate those of the rhombic dodeca-hedron, octahedron, cube, and tetrahedron. The icosahedral 31 belong to a different family of symmetries. Both groups together constitute seven sets of axes of symmetry: four contributed by the VE's vertices, edges, and two types of faces, and three by the icosahedron's vertices, edges, and faces.

1042.05 The seven unique cosmic axes of symmetry describe all of crystallography. They describe the all and only great circles foldable into bow ties, which may be reassembled to produce the seven, great-circle, spherical sets . . . .

We have a list of symmetrical possibilities. With this inventory, Fuller integrates specific information about inherent spatial characteristics with concepts of energy behavior, to gain insights about structuring in nature.

Excess of One

It is interesting to note that the number of great circles associated with each polyhedron is always one more than the number of its edges. For example:

	Edges	Great Circles
Tetrahedron	6	7
Octahedron	12	13
Cube	12	13
VE	24	25
Icosahedron	30	31

Fuller quickly identifies this constant "excess of one great circle" (and its implied two poles) with the "*excess two* polar vertices characterizing all topological systems" (1052.31). However, this discovery is not a new bit of magic, but rather follows directly from Euler's law. Recalling the way in which great circles are generated, we realize that each circle corresponds to a *pair* of either vertices, edges, or faces. Therefore, the number of great circles can be tallied by counting *half* the total number of topological aspects, or $\frac{1}{2}(E + F + V)$. We can now write an equation stating that one less than the number of great circles is equal to the number of edges:

$$\tfrac{1}{2}(E + F + V) - 1 = E.$$

Multiplying both sides by 2, we have

$$E + F + V - 2 = 2E,$$

or
$$F + V - 2 = E,$$
and finally
$$F + V = E + 2,$$
which is Euler's law.

Even if Fuller's cosmic railroad tracks leave you skeptical, great circles provide fascinating geometric patterns, which introduce a new system of classifying and comparing the topology and symmetry of various polyhedra. Part of their fascination lies in the surprisingly limited number of variations among the great-circle sets generated by our cast of polyhedra. But perhaps most importantly, experiments with great circles—building wire models—provided the impetus and the clues for Fuller's subsequent journey into geodesics.

From Geodesic to Tensegrity:
The Invisible Made Visible

"A geodesic is the most economical relationship between any two events" (702.01). Fuller's definition immediately calls to mind great circles, which provide the shortest routes between two events on a spherical system. Actually, clarifies Fuller, the general case of "most economical relationship" is necessarily a great circle. "It is a special case in geodesics which finds that a seemingly straight line is the shortest distance between any two points in a plane." In other words, a given area may be such a small portion of a spherical system that it appears flat; however, because all identifiable experiences belong to systems, "great-circle segment" and "geodesic" are interchangeable in synergetics.

Already familiar with the theory of great circles and their polyhedral symmetries, we can apply theory to practice. A little experimentation uncovers two important discoveries, demonstrating again why the discipline of building models was essential to Fuller's mathematical exploration.

The first discovery is easily visualized without actually building a model. Imagine a metal sphere and a wire ring just large enough to fit around the widest girth of the sphere. This circular ring qualifies as a great circle and therefore can delineate the shortest route between any two points on the sphere. However, there is a practical problem; the ring slides off the sphere. It may seem like a strange observation, but Fuller tended to investigate unusual aspects of his subject matter, and frequently such seemingly whimsical sidetracks have proved fruitful.

Bring in a second wire circle of the same diameter. As we recall from the last chapter, a pair of great circles intersect at two points 180 degrees apart. To keep the two wire loops on the sphere, they are tied together at both crossing points. The effort fails, for the two circles are free to spin around their common axis, and as soon as they line up both circles slide off the sphere together. Try again. A third great circle is placed anywhere on the sphere except through

Fig. 15-1

the intersection of the other two. Whether arranged randomly or symmetrically (as a spherical octahedron), when intersections are tied together all three circles are immobilized. Triangulation creates a stable cage. "Not until we have *three* noncommonly polarized, great-circle bands providing omnitriangulation... do we have the great circles acting structurally to self-interstabilize... " (706.20).

Three differently oriented great circles is the minimum for a stable model. The discovery may seem trivial at first, but we have no indication from design that the stability of a "three-way grid" is widely understood in our culture. On the other hand, Southeast Asians have utilized this principle for thousands of years. Fuller points out that a vital need for strong baskets led them long ago to discover that a triangulated weave stays rigidly in place, whereas the two-way weave used by other cultures is easily distorted.[1] Southeast Asian children today still play with a reed sphere consisting of six interwoven great circles—perhaps the oldest known toy (Fig. 15-1). Characterized by icosahedral symmetry, the ancient design utilizes the inherent stability of a three-way grid to make a lightweight and virtually indestructible ball out of delicate reeds. A modern toy has yet to improve upon its simplicity and durability.

The second discovery, which should be experienced to be fully appreciated, will nevertheless not come as a surprise at this point. Experimenting with wire models, Fuller found that the more great circles, the stronger the sphere. That much is self-evident, but the degree to which their strength increased far exceeded his expectations. This was not the kind of linear relationship exhibited by ordinary structures; rather the increasing rigidity of his great-circle models (from the minimum three to the icosahedral 31) could only be called synergetic.

We skip directly to the strongest model, provided by last chapter's limit case, 31 great circles. Triangulated geodesic arcs produce an

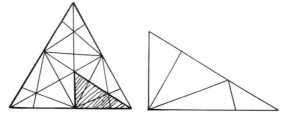

Fig. 15-2. LCD of the icosahedron is asymmetrically subdivided by the 31-great-circle pattern.

extremely sturdy, lightweight enclosure out of thin wire. Let's look closely at the pattern; our study is made considerably less complicated by isolating the LCD as described in the preceding chapter. Accordingly, we study only one of the 120 triangles framed by the icosahedron's 15 great circles, and observe that it is asymmetrically subdivided by the 6- and 12-great-circle patterns (Fig. 15-2). Notice the variations in arc length and surface angle—made more obvious by viewing this small region out of context. The overall three-way grid incorporates longer and shorter arcs, thereby subdividing the sphere's surface into triangles of very different shape and size.

The load distribution and resulting strength of these wire models is a function of symmetry; the longer the arc, the more vulnerable it is to stress.[2] It is clear that the most advantageous system would have all arcs as close to the same length as possible, but how can we improve upon the symmetry of the 31-circle pattern?

A second problem with the arrangement is that as a limit case, it does not present a logical course for further subdivision. To build progressively larger models with sufficient strength, we must find a way to generate more and more great-circle segments, or "higher frequency" in synergetics terminology. Both problems are solved by Fuller's next step.

To explore other methods of developing large multifaceted enclosures, we go back to the system that already has the greatest number of equivalent regular faces, the icosahedron. A three-way grid of evenly spaced lines (imagine a triangular checkerboard) divides the icosahedron's equilateral triangles into as many smaller triangles as desired (Fig. 15-3). However, this subdivision does not yet lead to an effective design strategy, for if neighboring triangles of a structural system lie in the same plane, the enclosure will deflect in and out, like a trampoline, in reaction to an applied load. Unless vertices are reinforced using rigid joints—in which case, the system

Fig. 15-3. Four-frequency (4v) triangles superimposed on each face of the icosahedron.

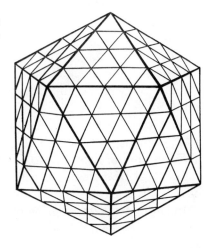

is functionally equivalent to the original icosahedron—the advantages of triangulation are lost. It is thus clear that a convex polyhedral enclosure with more than twenty triangles cannot consist exclusively of equilateral faces. Adjacent triangles must differ slightly in shape and size to allow angles around each six-valent vertex to add up to less than 360 degrees, as necessary for continuous convexity.[3] Fortunately, this unavoidable variation among chord lengths can be far less than that of the 31-great-circle pattern. To understand how these irregular triangles are generated, we back up and review the problem as a whole.

Theory Behind Geodesic Structures: Summary

We can think of Fuller's task as a mathematical game confined to the rules of synergetics, in which the goal is to enclose as much volume as possible with the least amount of material. A solution must combine two geometric principles—integrating overall shape considerations with the requirements for local stability. Geometry tells us that the shape with the greatest volume/surface-area ratio is a sphere; synergetics challenges that sphere to materialize. "Since physics has found no continuums, we have had to clear up what we mean by a sphere," writes Fuller (1023.11). This game calls for a solution we can actually build and touch, which means the best structural approximation of that elusive sphere.

We thus have redefined the problem in terms of "operational mathematics." To get started, we might experiment with toothpicks, thus limiting the game to enclosures with identical struts. This brings us back to Fuller's "three prime structural systems in Universe"—the inventory of self-stabilizing systems with equal vectors. Of the three —tetrahedron, octahedron, and icosahedron—the third has the most volume per toothpick (Chapter 5). An icosahedron is thus the best we can do with toothpicks, but of course it is not a satisfactory solution.

How can the enclosure become more spherical and still be stable? The geometric logic continues: through systematic symmetrical sub-division of our best approximation—taking full advantage of the stability of triangles. It hardly needs to be restated: "If we want to have a structure, we have to have triangles" (610.12).

In short, two simple principles taken together lead to a solution to the stated problem. Combine the advantageous shape of the icosa-hedron with the stability of triangles, and the geodesic dome almost materializes. The logic is as exquisite as it is simple.

But we're missing a step. How can we develop the "checkerboard" tessellation illustrated above into a functional structure? The icosa-hedral edges can be divided into any number of segments. The greater the number, or "frequency," the more spherelike the end result will be. Then, to complete the transformation, each new vertex (superimposed on icosahedron faces) must be projected out to the surface of the imaginary sphere defined by the icosahedral vertices—and then interconnected by great-circle chords (Fig. 15-4). The nature of this projection accounts for the slight variation in

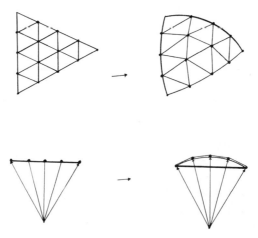

Fig. 15-4. Each new vertex is projected outward to the surface of an imaginary sphere defined by the original icosahedron vertices.

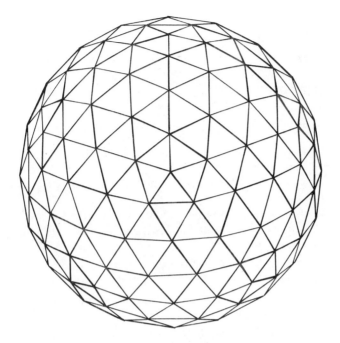

Fig. 15-5. 4v icosahedron: transformation of Figure 15-3.

shape and size of triangles; the farther away a chord is from an icosahedral vertex, the more it is stretched during the transformation from planar to spherical.

The resulting omnitriangulated polyhedron has six triangles at every vertex except for those located at the twelve vertices of the original icosahedron, which continue to join five triangles (Fig. 15-5). The system can have indefinitely many six-valent nodes, but the existence of exactly twelve five-valent vertices is a prerequisite to closure. Happily, there is also a considerable engineering advantage in having five or six struts leading out from each vertex: forces are instantly distributed in many directions—omniradially, Bucky might say—producing structures with unprecedented strength in relation to their weight.

This is the basic strategy behind the geodesic dome.

Geodesic Design in Nature

Fuller points out that an extremely high-frequency geodesic polyhedron provides the true model of physical systems which we interpret as spheres, as for example a soap bubble. The notion of a continuous surface equidistant from a central point is scientifically unacceptable, that is, inconsistent with physical reality; on some level of resolution all "spheres" consist of discrete quanta—untold numbers of energy events interconnected by an even greater number of vector-relationships, or forces. He has clarified this particular misconception countless times:

535.11 Because spherical sensations are produced by polyhedral arrays of interferences identified as points approximately equidistant from a point at the approximate center, and because the mass-attractive or -repulsive relationships of all points with all others are most economically shown by chords and not arcs, the spherical array of points produces... very-high-frequency, omnitriangulated geodesic structures... .

Our eyes cannot see individual molecules in the delicate transparent soap bubble, nor can we detect the chordal chemical attractions between molecules. Nevertheless they exist, explains Fuller, and it is our responsibility to understand and teach the truth about Universe. Once again, his goal is to provide tangible models of otherwise invisible phenomena.

Of all possible solutions, a high-frequency triangulated shell with icosahedral symmetry provides the most efficient method of enclosing space with a minimum of material and effort. Accordingly, nature relies on this elegant design in many situations calling for protective enclosures, regardless of scale. Examples include the small sea creatures called Radiolaria,[4] the fibrous web of the eye's cornea, and the protein shell of many viruses.[5] We looked into the structure of spherical viruses in Chapter 8 and can now go into greater detail based on our increased familiarity with geodesic theory.

The design problem is familiar by now: tiny amounts of genetic material must be protected by a tough protein shell. As nature is scrupulously efficient, the choice is clear. A "spherical" distribution of protein molecules will satisfy the basic criteria in terms of conserving material relative to volume, while icosahedral symmetry will provide the most even distribution. The natural balance sought by chemical forces leads to approximately equivalent spans, which is geometrically accomplished by high-frequency icosahedral systems. In short, the structure of viruses is a product of nature's eternal tendency toward equilibrium. Geometry imposes the rules.

Reassuringly, observations (with the electron microscope) of isometric virus shells have consistently revealed icosahedral designs. Dr. Aaron Klug, who first observed the geodesic structuring of viruses, wrote to Fuller in 1962 telling him of this discovery. Bucky, delighted by the news, immediately wrote back with the formula for the number of nodes on a shell ($10f^2 + 2$, varying according to frequency) as confirmation of Klug's hypothesis. Klug answered that the values obtained from this equation proved consistent with the virus research, and thereby provided Fuller with one of his most valued anecdotes—a prime example of nature's economic elegance —which enriched many lectures in subsequent years.

Insufficient awareness of spatial constraints causes these structural similarities to be perceived as "coincidence." As geodesic domes were utilized worldwide 15 years before electron microscopy enabled detection of virus capsids in 1962, the resemblance of the tiny biological forms to large architectural structures seemed to many quite extraordinary and improbable. Science does not as a rule take into consideration the active role of space; however, such awareness can assist the prediction of unknown structures based on their functional demands.

Another fascinating example was contributed to the inventory of geodesic structuring by scientists at General Dynamics working on the problem of rocket reentry, who wrote to Fuller describing their results and enclosing photographs. The experiment involved two hemispheres of thin-sheet titanium, precisely machined to achieve consistent shell thickness. The diameter of one was exactly an inch greater than that of the other, so that when the larger was placed over the smaller, a half-inch hemispherical cavity separated the two shells. Their bases were sealed together to create a double-shell dome, and the air was then pumped out of the intervening space to create a vacuum. Atmospheric pressure outside the dome caused its thin titanium sheet to buckle in toward the vacuum. The hemisphere "dimpled" in a "pure icosahedral pattern," as Fuller recalls in a 1975 lecture. Like the virus, it had no choice! The titanium sheet experienced an automatic reaction based on the shape of space; caving in most efficiently required a symmetrical distribution of dimples. The frequency of this pattern, that is, the number of dimples per icosahedral "edge," was found to be inversely proportional to the shell thickness: a thicker shell produced fewer dimples and vice versa.

Consider the above progression. We started with a mathematical puzzle; geometry laid out the rules and led to a solution, and it turned out that nature had been playing the same game all along. That is essentially how Fuller describes his experience in developing the geodesic dome:

> I did not copy nature's structural patterns. ... I began to explore structure and develop it in pure mathematical principle, out of which the patterns emerged in pure principle and developed themselves in pure principle. I then...applied them to practical tasks. The reappearance of [geodesic] structures in scientists' findings at

various levels of inquiry confirms the mathematical coordinating system employed by nature. (203.09)

The principles behind the geodesic dome are not new; they are eternal laws of nature. The application of these geometric facts to a building system *is* new. Fuller is quick to explain,

> Though...similar in patternings to...flies' eyes, geodesic structuring is true invention... Flies' eyes do not provide human-dwelling precedent or man-occupiable...structures. (640.01)

Invention can be defined as the novel application of generalized principles. Chapter 16 will explore the concept in more detail.

Geodesic Domes: Design Variables

The study of great-circle patterns seems at first like a completely abstract endeavor. However, the construction of models demonstrates that spheres with a greater number of shorter arc-segments have a significant structural advantage over simpler structures, and suggests potential applicability. Fuller's early models, based directly on great-circle patterns showed considerable strength but did not go far enough. A new method of generating geodesic structures was needed to produce higher-frequency structures—with less variation among chord lengths—than the 31-circle pattern. This progression led Fuller to concentrate years of attention on icosahedral geodesic designs. He discovered that the inherently self-stabilizing geodesic polyhedron could be truncated as the basis of a stable dome (Fig. 15-6). These domes, which could be any fraction of a geodesic sphere, must sit on the ground to complete a "system." Depending on the situation, a half-sphere truncation, $\frac{3}{4}$, $\frac{1}{8}$, or any number of other fractions might be desirable. Fuller saw that the possibilities were endless; geodesic domes could be designed to be built out of almost any material at any frequency.

The geodesic *pattern* is also a variable. In addition to the above "checkerboard," Fuller developed a number of other "geodesic

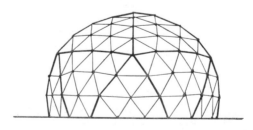

Fig. 15-6. 5/8th truncation of 4v icosa—as basis of a geodesic dome.

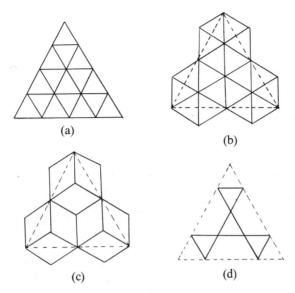

Fig. 15-7

breakdowns." An icosahedral face can be subdivided by a variety of patterns—each one yielding a different design for potential domes. The most common breakdown subdivides each triangular face with lines parallel to its three edges as explained above (Fig. 15-7a). The number of segments along each icosahedral edge specifies the *frequency* of the resulting geodesic dome, and the number of triangles per icosa face will always be f^2 for an "f-frequency" structure. A geodesic dome might be 2- or 32-frequency depending on size and material; there's no inherent upper limit, but the exponentially increasing numbers of different strut types become prohibitively complicated at very high frequencies.

The second breakdown subdivides icosahedral faces into triangles with lines perpendicular to icosahedral edges, producing slightly fewer triangles on the overall structure than the parallel version (12 per face for a 4-frequency breakdown, as compared to 16) and is somewhat less symmetrical (Fig. 15-7b). Fuller's diamond pattern is yet another choice (Fig. 15-7c). This design is unstable unless constructed out of panels rather than struts. If the situation allows mass production of panels, the diamond pattern will have an advantage over the first breakdown because of its fewer different types of faces. A number of aluminum domes have been built according to this design.

Finally, the deliberate omission of certain chords in Figure 15-7a produces Fuller's "basket-weave" design, characterized by hexagons and pentagons entirely framed by triangles (Fig. 15-7d). This pattern lends itself to building domes out of bamboo for example: struts in a three-way weave are simply tied together at crossings.

There are many other potential designs. The field is as wide open as the number of ways to symmetrically subdivide an icosahedron with great-circle chords. The fundamental characteristics of the resulting enclosures are the same. Geometric principles are exploited to gain unprecedented structural efficiency, just as nature, in response to the interplay of physical forces and the constraints of space, produces icosahedral geodesic patterns for many enclosures.

Geodesic domes of virtually unlimited size can be built by increasing the frequency as needed. The mathematics behind this undertaking is cumbersome but not conceptually difficult. Calculations are simplified (or at least kept under control) by an understanding of symmetry: complete information for an entire dome of any frequency is contained within its LCD triangle. Using the formulae of spherical trigonometry, we can manipulate the central and surface angles of spherical triangles to derive values with which to obtain strut lengths. Struts are great-circle chords, and each one is subtended by a specific central angle (Fig. 15-8a).

Depending on construction methods and materials, we might strive to keep struts as close to the same length as possible, or instead try to develop triangles with maximally similar shapes, or work toward any number of other preferred solutions. Frequency is itself an important variable; greater numbers of shorter struts may be more efficient in terms of supporting loads, but construction is correspondingly more difficult. Such tradeoffs must be carefully weighed, and fortunately trigonometry allows enormous flexibility in terms of potential solutions. Taking advantage of symmetry, we are able to experiment with different strut lengths in the LCD triangle, thereby only manipulating angles and lengths for a very small portion of the whole system, while developing the mathematical specifications for an entire geodesic dome (Fig. 15-8b). Finally, geodesic domes can be elongated or pear-shaped, or (theoretically) even shaped like elephants. The constant curvature of a sphere produces the greatest strength, but these other options do exist and can be developed with no more than a pocket calculator and a lot of paper.

As a final note, it is important to realize that while the theory behind geodesic domes is strikingly simple—and previously contemplated by others before Fuller was granted U.S. Patent 2,682,235 in 1954—the actual translation from theory to practical structures involved fantastically intricate mathematical development. As with

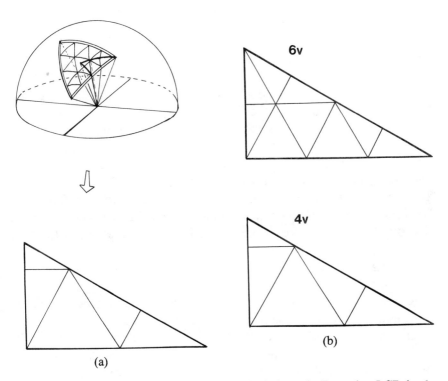

Fig. 15-8. (a) Use of LCD in geodesic-dome calculations. (b) Examples: LCD for $6v$ and $4v$ icosahedral geodesic domes.

most invention and design, the initial insight was not enough to produce a 250-foot-diameter clear-spanning structure overnight. The subsequent calculations required enormous aptitude and perseverance. Consider the precision necessary to have six struts meeting at the same point, at thousands of different vertices; minute errors in strut lengths at only a few points will accumulate and produce vast discrepancies elsewhere on the dome. Dealing with tolerances similar to that of the aircraft industry rather than the relatively crude building world, Fuller had to develop absolutely reliable trigonometric data to enable the construction of extremely large domes.

Tensegrity

Before geodesic domes appeared on the scene in 1948, the dome of St. Peter's Cathedral in Rome, with a diameter of 150 feet, was unchallenged as the largest architectural clear span. 150 feet must have been seen as a fundamental upper limit, a sort of divine zoning law. Ultimately, the inherent stability of triangles cannot alone account for the geodesic dome's ability to span unlimited distances with no interior supports—nor for its unprecedented strength-to-weight ratio.

The rest of the explanation lies in an understanding of *tensegrity*, Fuller's contraction of the two words *tension* and *integrity*. "Tensegrity describes a structural-relationship principle in which structural shape is guaranteed by the...continuous, tensional behaviors of the system and not by the discontinuous exclusively local compressional member behaviors" (700.011); Fuller thus introduces a discussion of the interplay of tension and compression forces in Universe. His term also refers to the inescapable co-occurrence of tension and compression, while its first syllable emphasizes the too often overlooked role of tension.

All systems consist of some combination of tension and compression forces. The two are inseparable.

All systems? What about a simple piece of nylon rope pulled at both ends between two hands? Isn't that pure tension?

Try to visualize the experiment: pull as hard as you can on both ends and notice what happens to the thickness of the rope. Its diameter shrinks slightly, betraying the invisible compression force around the rope's circumference, or perpendicular to the applied tension. Even in this simple example, unplanned compression is inevitable.

The same is true in reverse. Applying compression to a column introduces surface tension around its girth. Pushing from both sides along the long axis of a strut causes the outside surface to stretch, albeit slightly. Tension and compression go hand in hand, as simultaneous complementary functions; however one or the other usually dominates a given situation.

641.01 No tension member is innocent of compression, and no compression member is innocent of tension....

641.02 Tension and compression are inseparable and coordinate functions of structural systems, but one may be at its "high tide" aspect, i.e., most prominent phase, while the other is at low tide, or least prominent aspect. The *visibly* tensioned rope is compressively contracted in almost *invisible* increments of its girth dimensions.... This low-tide aspect of compression occurs in planes perpendular to its tensed axis....

We thereby draw the distinction: our rope is a tension element; the loaded column, a compression member.[6] Effective design must balance the two interdependent forces in preferred ways.

The apparently insurmountable limit to the clear-span of a structure existed because the interdependence of tension and compression was not fully understood. In general, the history of construction reveals an overwhelming dependence on compression. Our concept of building has been inseparably tied to that of weight; early humanity piled one stone on top of another, and we continue to employ the same single strategy, fighting gravity with sheer mass. But compressional continuity has its limits, such as the impossibility of achieving spans greater than 150 feet. Any larger dome would collapse under the force of its own weight. Moreover, although architects may not have reflected on the principles of tensional integrity, necessity apparently forced them to add a powerful iron chain around the base of St. Peter's dome; the outward thrusts from all that compression needed further restraint.

Fuller decided that a better approach was needed than that of slapping on a bandage at the end. Following nature's example, tension must be designed into the structure at the start. In fact, tension must be primary.

Nature's Example

Look around; nature's been using tensegrity all along. Humanity was able to overlook this structural truth for thousands of years because tension tends to be invisible. Seeing rocks sitting on the ground and bricks piled upon bricks, we have developed a virtually unshakable "solid-things" understanding of how Universe works. The ubiquitous tension forces, from gravity to intermolecular attraction, tend to be more subtle:

> ... at the invisible level of atomic structuring the coherence of the myriad atomic archipelagos of a "single" pebble's compressional mass is provided by comprehensively continuous tension. This fact was invisible to and unthought of by historical man up to yesterday... there was naught to disturb, challenge or dissolve his "solid things" thinking...[7]

The inseparable partnership of tension and compression does not mean that the two forces are the same. On the contrary, fundamental differences are the basis of their successful interdependence.

Compression is local, discontinuous, says Bucky. When we load a column, we push it together. If we push a thin column too hard, the column will buckle like a banana; there is no other way for the stressed member to yield. For this reason, compression members are subject to an inherent limit to their length relative to their cross-sectional area, called the "slenderness ratio." The development of stronger materials has increased that ratio only slightly over thousands of years: from a maximum of 18 : 1 for stone columns in ancient Greece to approximately 33 : 1 for modern steel. The limit is not as much a result of inadequate materials as of geometry: compression is directed inward, and hence eventually forces the overstressed column to buckle. Compression fights against the shape of a strut.

Tension, on the other hand, pulls apart. The direction of tension serves to reinforce the shape of the stressed member. Pulling straightens; pushing bends. As a result, assuming stronger and stronger materials, there is no inherent geometrical limit to the length of a tension component. While the capability of compression members has remained more or less the same, tension materials have improved by leaps and bounds, and significant advances continue today.

Compression was the sole basis of man-made structures until the tensile strength of wood was exploited, enabling for the first time the construction of structures light enough to float on water: simple rafts, initially, followed by progressively more sophisticated rowing and sailing vessels. But at 10,000 pounds per square inch, the tensile strength of wood was still overshadowed by the 50,000 psi compression strength of stone masonry. Not until 1851 saw the first mass production of steel—with a tensile strength equal to its compression-resisting capability of 50,000 psi—was tension finally brought into parity with compression, explains Fuller; "so tension is a very new thing." This development enabled the Brooklyn Bridge in 1883 and ushered in a whole new era of tensional design. Scientists rapidly created metal alloys of greater and greater tensile strength with less and less weight, ultimately leading to jet airplanes and other previously inconceivable miracles. A new material called carbon fiber, with an unprecedented 600,000 psi, was responsible for a recent "miracle". In 1979, Paul MacCready pedaled the first human-powered aircraft, which he called the "Gossamer Albatross," over the English Channel; two years later he repeated the journey with a completely solar-powered plane. The only reason he was able to accomplish this feat, emphasizes Fuller, was that the extraordinary tensile strength of carbon fiber allowed the plane to have a wing span of 96 feet, while weighing only 55 pounds; you could hold it up in one hand. But the newspapers didn't mention the carbon fiber, he declares sternly; "nobody talks about this invisible capability." This beautiful example of "doing more with less" was a favorite for Bucky, who never forgot being told by well-meaning adults during his first eight years, "Darling, it is inherently impossible for man to fly."

Driving toward a specific observation about Universe, Bucky describes theoretically ideal structural components for each of the two forces, as suggested by their different characteristics. Why will a short fat column not buckle under a compressive force that easily breaks a tall thin column of the same material and weight? Geometry governs the situation as follows. A system is most resistant to compression in one direction, namely along an axis perpendicular to its widest cross-section, in which case, its vulnerable girth is as strong as possible. This is the neutral axis. The wider its girth, the more impervious a column will be to compression. Therefore, a short fat column is better able to resist buckling than a tall thin column of the same material and weight, because the latter lacks sufficient resistence perpendicular to the line of force. Similarly, if a compression force that will easily break a long column (a pencil, for example) is applied perpendicular to its length, that column will be unharmed (Fig. 15-9a, b). That much is common sense, but in less extreme situations an understanding of the "neutral axis" is necessary to enable the prediction of exact results.

Next, imaging loading a slightly malleable cigar-shaped column; compression causes the girth to expand, forcing the column to become progressively more spherical. This transformation suggests a candidate for the ideal compression component (Fig. 15-9c).

A sphere is the only shape in which every axis is a neutral axis, which is to say, a sphere's width is the same in every orientation. Therefore, this shape resists compression from any direction; it cannot buckle. Hence the ball bearing. This tangible example illustrates the ideal design for compression.

What about tension? Evidence of longer, thinner, and ever more resilient tension materials suggests that there's no inherent limit to length. Fuller takes this a step further: "May we not get to where we have very great lengths and no cross-section at all?" He answers his own question, "This is just the way Universe is playing the game." Gravity *is* that invisible limitless tension force. "The Earth and the Moon are invisibly cohered...."; the tension cable has reached the limit case in thinness: it's nonexistent. "You have enormous tension with no section at all." A splendid design! The solar system is thus a magnificent tensegrity: discontinuous compression spheres (i.e.,

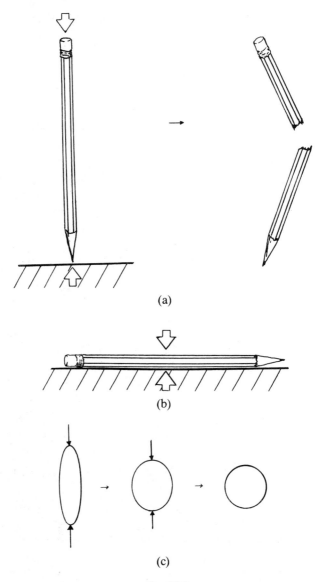

Fig. 15-9

planets) are intercoordinated—never touching each other—by a sea of continuous tension. "Every use of gravity is a use of . . . sectionless tensioning," Fuller continues, observing that "this is also true within the atoms: true in the macrocosm and true in the microcosm" (645.03–5).

By mid-twentieth century, it was clear that the design plan of Universe involves islanded compression and continuous tension, but up until then

> man had been superficially misled into [thinking] that there could be solids or continuous compression.... Only man's mentality has been wrong in trying to organize the idea of structure. (645.04)

New Concept of Construction

That humanity can learn from the principles of nature is the essence of Fuller's message. We must abandon our building-block concept of structure in favor of comprehensive solutions which take advantage of the inherent qualities of tension and compression. The latter tends to do the local, isolated structural tasks in nature, while the former specializes in cohering systems over great distances. While we understand that Universe is not structured like a stack of bricks, that awareness has not affected our approach to construction. A "building-block" approach has persisted more or less unchanged for thousands of years, pitting structural bulk against gravity's vigilant force. Instead, argues Fuller, we must think in terms of whole systems in equilibrium, omnidirectional forces interacting in self-stabilizing patterns. If, emulating nature, structural design capitalizes on the integrity of tension, these "whole systems" will prove far stronger than analysis of their separate parts could predict.

Additionally, we can learn from nature's structuring method: converging and diverging, she produces bubbles, explosions, stars, and the radially expanding sound and light waves. Energy pushes out, and its expansion is countered by tensional restraints such as the pull of gravity and molecular forces. Eventually, the two dynamics reach an equilibrium, a tentative balance such as a soap bubble.

Fuller heralded an "era of thinking and conscious designing in terms of comprehensive tension and discontinuous compression" (640.42). He saw an unmistakable change taking place, largely going unnoticed. This "new era" began with the spoke wheel, which Fuller pinpoints as man's first breakthrough into tensegrity thinking:

> I saw that his structural conceptioning of the wire wheel documented his intellectual designing breakthrough into such thinking and structuring. The compressional hub of the wire wheel is clearly islanded or isolated from the compressional "atoll" comprising the rim of the wheel. The compressional islands are interpositioned in structural stability only by the tensional spokes.... This reverses the historical structural strategy of man. (640.42)

The wheel's use of tension enables a far more efficient and lightweight

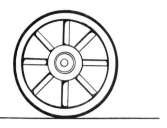

Fig. 15-10. Wheel with compression spokes.

structure than could be produced with compression spokes. Tension materials are inherently smaller and lighter than compression materials carrying equivalent loads.

The wheel was originally an exclusively compression structure—starting with the cave man's stone cylinder and progressing to slightly more sophisticated designs like "the old artillery wheel" cited by Fuller in *Synergetics*.[8] It continues to be perceived as such. (A version of that wheel is sketched in Figure 15-10.) Widespread awareness of the tensional integrity responsible for the spoke wheel's lightweight efficiency has not been reached. The load on a bicycle wheel is therefore often seen as "sitting" on the lower spokes—like columns—rather than hanging from the top spokes. Despite design breakthroughs, humanity as a whole is still caught in "solid-things" thinking.

Many other structures that rely on tensional integrity can be cited, such as suspension bridges and sailboats. However, Fuller points out that tension was usually incorporated as a "secondary accessory of primary compressional structuring." In other words, ancient man—habitually relying on compressional continuity—inserted a "solid" mast into his hull, but finding that the wind kept blowing his mast over, he added a set of stabilizing tension wires, or *stays*, in nautical terminology. Somehow he failed to learn from his accidental design, in which the extraordinarily thin, lightweight tension adjuncts withstand the same forces as the heavy solid mast.

Tension has been secondary in all man's building and compression has been primary, for he always thought of compression as solid. ... Earth and ship seemed alike, compressionally continuous. (640.50)

Modeling the Invisible

Tensional integrity is certainly how Universe works, pondered Bucky in the 1940s, but how can I illustrate this invisible phenomenon? Can the co-occurrence of discontinuous compression and continuous tension be modeled in such a way as to bring this structural principle into easy grasp? He was determined to display the invisible truths of science in a scale that can be perceived by human senses.

Could it be done? Could discontinuous reality be modeled? Science in the twentieth century feels exempt from modelability, philosophized Bucky; ever since the isolation of the electron in 1898, scientists have felt increasingly more sure of their lack of responsibility to explain their work to the layman. The hypothetical scientist of Fuller's lectures declares, "I am sorry to say, reality is both invisible and unmodelable."

In the summers of 1947 and 1948, Fuller taught at Black Mountain College, and spoke constantly of "tensional integrity." Universe seems to rely on continuous tension to embrace islanded compression elements, he mused; we must find a way to model this structural principle. Much to his delight, a student and later well-known sculptor, Kenneth Snelson, provided the answer. He presented his discovery to Fuller: a small structure consisting of three separated struts held rigidly in place with a few strings. This was the birth of an explosion of geometric tensegrity structures. Inspired by Snelson's discovery, Fuller went on to create tensegrity versions of countless polyhedra.

Tensegrity Polyhedra

Tensegrities can be derived from all polyhedra, whether regular, semiregular, high-frequency geodesic, or irregular, typically with one strut representing each edge of the polyhedron. Struts do not come in contact with each other, but instead are held in place by a network of tension elements, or strings—producing completely stable sculptural systems. The complexity of molecular interactions aside, the two types of components are characterized by axial-force states, thereby using materials most efficiently, because components can be far lighter than would be required to withstand bending. The technical sound of the words in no way prepares you for the exquisite appearance of these structures. Photograph 15-1 shows a tensegrity icosahedron and tetrahedron, while Photograph 15-2 displays a $3v$ tensegrity icosahedron, as representative tensegrity polyhedra. However, there's no substitute for actual models.

At last, says Bucky, we are able to experience at first hand the truth about structure: systems cohere through tensional continuity, and nothing in Universe touches anything else:

...Tensegrity structures satisfy our conceptual requirement that we may not have two events passing through the same point at the same time. Vectors [i.e.,

Photo. 15-1. Tensegrity icosahedron and tensegrity tetrahedron. Photograph by Amy C. Edmondson.

struts] converge in tensegrity, but they never actually get together; they only get into critical proximities and twist by each other. (716.11)

A tensegrity icosahedron, therefore, is more honest than a toothpick–marshmallow structure which *seems* to have five edges *touching* at each vertex. In the tensegrity, edges all come within critical proximity of the location of a "vertex" and "twist by each other." Figure 15-11 illustrates the relationship between the tensegrity icosahedron and its Platonic counterpart. Instead of a five-valent "point" with the illusion of continuity, each convergence is marked by a pentagon of string, thus illustrating the fact that individual energy events do not touch but instead hover in a state of dynamic equilibrium. Tensegrity structures provide a visible, tangible illustration of an invisible truth. Forces and their interactions are brought out in the open.

Perhaps the most significant lesson from tensegrity structures lies in their unexpected strength. A tensegrity's apparent extreme fragility is completely deceptive. The uninformed observer will usually approach such a structure with great caution, touch it hesitantly and gently so as not to break the delicate model, and (if persistent) ultimately realize that a tensegrity can be thrown around the room without harm. The erroneous initial assumption is a result of a deeply ingrained bias in favor of compression as the reliable source of structure. We perceive string and cable as flimsy, but actually the

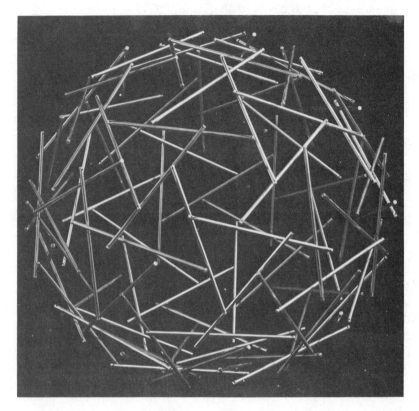

Photo. 15-2. $3v$ tensegrity icosahedron with 90 struts. Photograph courtesy of Thomas T. K. Zung, Buckminster Fuller, Sadao and Zung Architects, Cleveland, Ohio.

magnitude of a force that can be carried by "delicate" tension materials can far exceed the corresponding capacity of compression elements. Humanity's perception is in need of retuning.

The other aspect of a tensegrity's remarkable strength is the rapid omnidirectional distribution of applied forces. One of the advantages of a network of tension elements is efficient dispersal of loads around a structure, enabling the whole system to withstand forces far greater than could be predicted by engineering analysis of the separate components. Welcome back, synergy:

> This is not the behavior we are used to in any structures of previous experiences... Ordinary beams deflect locally... The tensegrity "beam" does not act independently but acts only in concert with "the whole building", which contracts only symmetrically when the beam is loaded.... The tensegrity system is synergetic. ... (724.33–4)

This "whole system" behavior can be detected by pushing or pulling

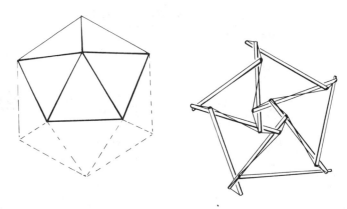

Fig. 15-11. Relationship of tensegrity icosahedron to its Platonic (planar) counterpart.

on two opposite struts of certain tensegrities. The entire system will contract or expand symmetrically like a balloon, and also will spring back to its equilibrium configuration when the applied force is removed. (Photo 15-3 shows the simple six-strut tensegrity, which is arguably the most elegant illustration of this uniform contraction or expansion in response to a unidirectional force.) Similarly, this synergetic behavior insures a balanced distribution of stresses:

> If you... tauten one point in a tensegrity system, all the other parts of it tighten evenly. If you twang any tension member anywhere in the structure, it will give the same resonant note as the others. (720.10)

Fuller felt that this dispersal is not really understood by most engineers, and as a result, they have been unable to predict or analyze the extraordinary capabilities of tensegrities and geodesic domes. Note that discontinuous compression and continuous tension also characterize geodesic domes, but as these structures lack the visually legible quality of the geometric models, one cannot at a glance identify the operative forces:

> Structural analysis and engineering-design strategies... were predicated upon the stress analysis of individual beams, columns, and cantilevers as separate components... [and] could in no way predict, let alone rely upon, the synergetic behaviors of geodesics.... Engineering was, therefore, and as yet is, utterly unable to analyze effectively and correctly tensegrity geodesic structural spheres in which none of the compression members ever touch one another and only the tension is continuous. (640.02)

The ultimate result of this conceptualizing and model building is that the barrier to ever-larger clear-spanning enclosures has been

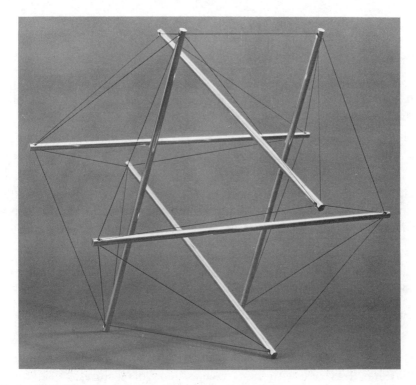

Photo. 15-3. Six-strut "expanded octahedron" tensegrity. Photograph courtesy of the Buckminster Fuller Institute, Los Angeles, Calif.

removed. By understanding the crucial role of tension, we can learn to manipulate it in preferred ways. "We are able to reach unlimited spans because our only limitation is tension, where there is no inherent limit to cross-section due to length" (764.02). Cosmic zoning laws are repealed.

Pneumatics

We recall Fuller's great-circle description in which a vast number of gas molecules are bouncing around inside a sphere, with their great-circle chords ultimately describing an icosahedral pattern as a result of spatial constraints. Tensegrity now completes the image.

Fuller explains that tensegrity provides a tangible demonstration of what happens inside a balloon. We tend to think of the balloon's skin as a continuous surface; however, a more accurate picture is that of a network of molecules in close proximity, such that the spaces between them are smaller than air molecules, allowing the network to act as an effective cage:

> The balloon is indeed not only full of holes, but it is in fact utterly discontinuous. It is a net and not a bag. In fact, it is a spherical galaxy of critically neighboring energy events. (761.03)

Gas molecules push out against the tensed rubber net and a dynamic equilibrium is maintained; compression and tension are in balance.

Our eyes cannot see the bustling molecular activity of the balloon, and Fuller sees synergetics as a way to help us tune in to this invisible behavior. A high-frequency tensegrity icosahedron does just that:

> In the geodesic tensegrity sphere, each of the entirely independent, compressional chord struts represents two oppositely directed and force-paired molecules. The tensegrity compressional chords do not touch one another. They operate independently, trying to escape outwardly from the sphere, but are held in by the spherical-tensional integrity's closed network system of great-circle connectors.... (703.16)

He elaborates on the parallel, holding up his icosahedral tensegrity that has traveled with him to hundreds of lectures: "This *is* a balloon, except that the tension components are only placed right where they're needed."[9] It's a balloon with all the excess tension taken out; strings are located only where the strut (molecule) wants to impinge on the sphere's surface.

We are thus led back to the necessity of three-way great-circling:

> A gas-filled balloon is not stratified. If it were, it would collapse like a Japanese lantern.... Once we have three or more...push-pull paths [of paired kinetic molecules] they must inherently triangulate by push-pull into stabilization of opposite angles. Triangulation means self-stabilizing; which creates omnidirectional symmetry; which makes an inherent three-way spherical symmetry grid; which is the geodesic structure. (766.02–4)

The analogy is complete. Pneumatics are dynamic high-frequency tensegrity geodesic configurations.

The principle of tensegrity, perhaps more than any other single aspect of synergetics, has yet to be exploited to its real potential in terms of design advantage. But the elegant simplicity of these remarkable structures hints at the nature of a future design revolution—toward innovative designs with unprecedented performance per pound. Finally, the tensegrity system, besides suggesting structural applications, has also provided a useful model in science.

Case in Point: Donald Ingber

One striking example was contributed to the growing list in 1983 by Donald Ingber, then a doctoral student at Yale University working on the biology of tumor formation and malignant invasion. Ingber had been exposed to tensegrity structures in an undergraduate design course—a playful option on the other end of the academic spectrum, which was to influence his vision as a scientist in profound ways. While pursuing his research in cell biology, Ingber began to observe some fundamental similarities between the subjects of his two seemingly opposite investigations. It appeared that tensegrity theory was applicable to biological systems; that is, Fuller's structures exhibited certain dynamic characteristics that were analogous to cell and tissue behavior. Significant structural parallels between the behavior of cellular and tensegrity systems led Ingber to powerful insights about the regulation of cell shape, differentiation, and growth, and thereby suggested a strategy for further investigation.[10]

Ingber proceeded to build, test, and study tensegrity structures in an effort to understand the implications of his proposed model, and was excited by the results of the comparison. His revolutionary approach has led to significant breakthroughs in his research into cancer formation, which he now continues at Harvard University Medical School. Without a detailed description of Ingber's research, we can still profit by the theme of his radical theory. He proposes a "whole system" approach in which "the architectural form of a tissue may itself serve to coordinate and regulate the shape, orientation, and growth of its individual cells through transmission of the physical forces of tension and compression characteristic of a given three dimensional configuration."[11]

There is no more appropriate conclusion than that provided by Ingber's own description in his 1983 letter to Fuller:

The beauty of life is once again that of geometry with spatial constraints as the only unifying principle. It is of interest to note that, as presented in the accompanying paper, cancer may then be viewed as the opposite of life resulting from a breakdown of this geometric hierarchy of synergetic arrangements.[11]

"Design Science"

"I did not set out to design a geodesic dome," Fuller once said, "I set out to discover the principles operative in Universe. For all I knew, this could have led to a pair of flying slippers." This playful declaration stands as a concise summary of the philosophy behind Fuller's life's work and introduces the relationship of synergetics to design. "Design science," in the most general terms, maintains that faithful observation of Universe is the basis of successful invention. The idea therefore is not to invent some strange new gadget, hoping there will be a market for it, but rather to tap into the exquisite workings of nature. While the significance of scientific discoveries is not always immediately understood, the accumulated "generalized principles" have been applied in innovative ways throughout history, producing artifacts which have gradually transformed the physical environment. Therein lies the key to humanity's success aboard Spaceship Earth, explains Bucky Fuller.

"Comprehensive Anticipatory Design Science"

Characteristically, his title expands—to embrace the full significance of this vital human endeavor.

Fuller defines *design* as the *deliberate ordering of components*. Thus distinguished from randomness, design implies the presence of intellect. His definition is worth our serious attention, for the word is too often associated with the concept of decoration—secondary or superficial embellishment as opposed to thoroughly developed systems. Fuller was quick to point out that Universe overflows with evidence of design; unimaginably intricate and reliable energy patterns reveal "eternal design interrelationship principles." Technology, to Fuller, is principle in action, and so "Universe is nothing but incredible technology." Its awesome complexity is the inspiration for Fuller's phrase "Intellectual Integrity of Eternally Regenerative Universe"—a weighty title attempting to convey a nonanthropomorphic respect for a greater (in fact all-encompassing) divine intelligence.

Combining this newly defined word with "science," to describe a new discipline or field, further enriches its significance. "Science" hints at the necessary rigor, suggesting a systematic new study. Fuller thereby expands the realm of "design"; the scientific method is essential, for "design science" involves the application of principle. He points out that "generalized principles" are eternal truths, as opposed to special-case statements or transient facts, and as such are inherent aspects of reality waiting to be discovered. Only human beings are able to discern such truths (science) and thereby participate in their own evolution (design). "Design science" is thus saturated with meaning: humanity alone has access to the design laws of Universe, and that has determined our unique evolutionary function. Just as bees are meant to cross-pollinate, we are meant to solve problems. Without specialized long beaks or wings or other role-specific physical traits, human beings have learned to exploit mechanical advantage, discipline the electron, travel more quickly than the fastest leopard, and fly farther than the strongest bird. Our unique advantage is a faculty called "mind," which can integrate disparate facts of experience. We are therefore deliberately designed to be "comprehensivists" while all other creatures are specialists. This brings us to the next aspect of Fuller's wordy title.

"Comprehensive..."

Fuller was profoundly impressed by the danger of overspecialization. He was once asked to speak at a convention of the American Association for the Advancement of Science, and the experience provided one of his best parables. Whether by luck or through Fuller's characteristic genius for detecting significant patterns, he happened to encounter two papers with a striking similarity—presented at different sections of the conference. The reports, on biology and on anthropology, both happened to discuss the phenomenon of extinction: the former investigating various extinct species, and the latter, extinct human tribes. Both papers concluded that the cause of extinction was overspecialization, which, taken to an extreme, precludes general adaptability. Fuller took the message to heart.

He says his predilection for thinking comprehensively began with his World War I Navy experience. Belonging to the last generation of sailors that preceded ship-to-shore radiotelephone equipment,

officers in 1917 were still trained as "comprehensivists." Everyone
had to be capable of handling any job on the ship, for voice
communication with land was not yet possible. The need for all
members of a crew to act quickly in an emergency demanded quick
"comprehensive" thinking and the flexibility to take over any job
without instructions from superior naval powers. A second lesson
was that a sailing vessel is itself a managable whole system, and
every member of her crew is working toward the success of the whole
ship. It seemed to Bucky a very desirable way to operate. He began
to see that our entire planet is one system and deliberately set out to
understand the interrelatedness of human affairs. He later coined the
famous "Spaceship Earth" to symbolize this approach, as it became
ever more apparent that effective design had to recognize the existence of a finite and inescapably connected whole-system world.
Lack of such awareness leaves us stranded on a ship with the
starboard side short-sightedly using much of its time and energy in
an effort to sink the port side, and vice versa.

Fuller then calls our attention to a subtle irony implicit in the
compartmentalization of the sciences, which require an ever narrower focus as one pursues a given scientific discipline more deeply,
as if to deny the relatedness of various aspects of scientific knowledge. At the beginning of this century, he recalls, chemistry and
biology, for example, were totally separate fields, the former encompassing chemical elements and their reactions, the latter pondering
the classification of species and the mystery of evolution. Moreover,
all branches of science involved philosophical speculation. Time was
given to questions such as "what is life?" and the difference between
life and nonlife could not have seemed more self-evident.

In this century, however, science faced an unpredicted development. As a necessary reaction to new discovery, new fields emerged,
such as biochemistry, defying the rigid boundaries between disciplines. (Science responded by making individuals that much more
specialized within such new categories, laments Fuller.) More
sophisticated equipment had revealed the chemistry of life, forcing
scientists to integrate their fields. The helix of nucleotides in DNA,
the magnificent chemical transitions in photosynthesis, and hundreds
of newly observed reactions belonged inarguably to both chemistry
and biology, and precipitated the birth of "biochemistry." And
gradually the clear boundary between life and nonlife was dissolving,
for both consist ultimately of electrochemical process. But scientists
were no longer the "natural philosophers" characteristic of the turn
of the century, maintains Fuller, and so this astonishing evolution

went largely unheralded. Specialization removed the burden of asking what a scientific development *means* in terms of the whole picture.

Fuller has a reason for pointing out such historical trends: the design scientist will be maximally effective as a comprehensive thinker. Once again, he redefines and thereby expands the concept of design. The design scientist is not to be concerned with an attractive handle for refrigerator doors, but rather with the whole concept of the distribution and preservation of food for humanity. Such subjects, he emphasizes, are *not* too large to think about. Only through systematic comprehensive planning does humanity have a chance to survive its growing crisis.

A "comprehensivist," he continues, may periodically have to "plunge very deeply" into a narrow subject or specialized project; however, such activity is always part of a larger plan. "Local problem solvers" can function with a global perspective. Our emphasis must shift from "earning a living" to accomplishing vital tasks if humanity is to survive, cautions Fuller; moreover, the "living" will take care of itself if we concentrate on doing what needs to be done.

These sweeping statements at first may seem difficult to apply; however, Fuller's philosophy is backed up by a lifetime of revolutionary invention and research into world patterns and trends, which stem directly from his 1927 decision to think about the whole system of "Spaceship Earth." He explains that his accomplishments were only possible because he gave himself the license to be a generalist: it *is* feasible to conceptualize humanity's food production and distribution (for example) as a whole system; it is a complicated study, but one with clearly defined boundaries: just food—where it is grown, where and when it is eaten, how and at what cost to the environment, consumer, etc.[1]

"...Anticipatory..."

Finally, the design scientist must think ahead. In each industry, there are specific "gestation rates" that determine the length of time between invention and widespread practical application. These inherent lags vary according to the nature of a design: in the electronics industry, for example, it is only a matter of months before a new invention can be incorporated into commercial production; car manufacturers might require five years to bring a new idea to the consumer; and housing presents the slowest evolution of all, Fuller's rather optimistic estimate of the 'gestation rate' being fifty years.

Psychological resistance to change, absence of urgency, and ignorance keep our approach to housing many generations behind our technological capability. The design scientist must take these lags into consideration, explains Fuller; an invention often must wait until its time, but the designer has a responsibility to anticipate long-term developments. Finally, a necessary implication is that we can glean important clues through the study of trends, and thereby determine what needs to be done.

So let us look at the relationship between "comprehensive anticipatory design science" and synergetics. Invention, as stated above, is the novel application of one or more "generalized principles." In the previous chapter, we discussed two inventions, the geodesic dome and tensegrity structures, in light of this statement. Briefly, the dome combines the inherent stability of triangles with the advantageous volume-to-surface-area ratio of spheres. A variety of structures based on geodesic chords could satisfy the above requirements, but icosahedral symmetry approximates the spherical distribution most efficiently. Both geodesic domes and tensegrities are direct applications of the principle describing the specific interdependence of tension and compression.

The theory of geodesic domes is taken a step further by an additional principle: the varying rates of geometric expansion. As discovered and exploited long ago by clipper-ship owners—as well as by today's shipping industry—a ship with twice the length of another has eight times the volume and four times the surface area. Translated into practical advantage, the cargo (i.e. payload) of larger vessels increases rapidly with respect to the amount of material and drag, which together determine the effort and cost of building ships and driving them through the sea. Larger ships are therefore more cost-effective. Despite the pivotal role of this geometric principle in shaping the historical direction of shipping, the flow of resources, and the ultimate mobility of humankind, these varying rates are not popularly recognized. With respect to geodesic domes, this means doubling the diameter increases the material fourfold and encloses eight times the volume. As both cost and temperature control of an enclosure are directly determined by surface area, the efficiency of a geodesic dome increases drastically with size. The implication is that their true design-science advantage is yet to be realized and may involve very large structures in novel environment control applications. Whether utilized for shelter, food production, recreation, or other functions, geodesic domes can enclose so much space with so little material that unprecedented future applications are not unlikely. In conclusion, the geodesic dome embodies design science at work.

Dymaxion Map[2]

Another of Fuller's inventions, in response to a very different problem than architectural design, is based on similar geometric principles. The problem is to draw a flat map of the world without the gross distortions inherent in the Mercator projection. In the early 1940s, dismayed by this widely accepted map's inaccurate depiction of our world—a visual lie presenting Greenland as three times the size of Australia, when exactly the reverse is true—Fuller was determined to discover a better solution.

Let's consider the design problem. Visual data must be reliably translated from the surface of a spherical "whole system" onto a flat display with only one side. To understand how the Mercator projection attempts to accomplish this task, imagine wrapping a large rectangular piece of paper around a transparent globe, forming a cylinder that touches only the equator. The geographical outlines are then projected directly outward to the cylindrical paper, as if by a light source inside the globe casting omnidirectional shadows. As a result, the visual information is accurately translated to the paper only at the equator; some distortion exists slightly above and below the equator, and it increases radically as one goes farther north and south on the map. In many versions Antartica is left out altogether, even South America is smaller than the gigantically distorted Greenland, and certain land areas—depending on which country has produced the map—must be split in half to turn the cylinder into a flat poster.

Unwilling to accept such distortion as necessary, Fuller started from scratch. If a spherical system is to be translated onto a flat surface, what is the most efficient and direct solution? It's a geometry problem; the relevant "generalized principle" involves the polyhedral system which best approximates a sphere with only one type of face. (The latter consideration insures an evenly distributed projection.) That system, which is of course an icosahedron, is the basis of a reliable and simple solution.

Imagine a globe with the edges of a spherical icosahedron superimposed on its surface by thin steel straps. Chapter 14 described planar polyhedra expanding into spherical polyhedra, as if drawn on balloons; we now visualize the reverse process. The steel arcs slowly unbend into straight-edge chords, while the curved triangle faces flatten out into planar equilateral triangles. The overall shape change is relatively slight (consider for comparison a spherical tetrahedron undergoing the same operation, or a sphere turning into a cylinder as in the Mercator projection), and the global "whole system" is preserved. As the globe transforms into an icosahedron, twenty spherical triangles with 72-degree corners become planar triangles with 60-degree corners; 12 degrees are squeezed out of each angle. (With three angles per triangle, 12 degrees times 60 angles equals none other than our old friend the "720-degree takeout.") Because each triangle of the spherical icosahedron covers a relatively small portion of the sphere and is thus fairly flat, the distortion during this transformation is minimal—and in fact invisible to the untrained eye. Moreover, the polyhedral projection automatically distributes the distortion symmetrically around the globe's surface and thereby insures that the *relative* sizes of land masses are accurate. (This is why a regular polyhedron is a preferred vehicle; different types of faces would distort slightly different amounts during the transition from spherical to planar.) Finally, all geographical data are contained within the triangular boundaries; there is no "spilling" of information or need to fill in gaps with "extra" land as in the Mercator.

The next step is straightforward. Unfold the icosahedron to display its twenty triangles on a flat surface. The result is a map of the entire world with little distortion of the relative shape and size of land masses, and no breaks in the continental contours (Fig. 16-1). No nation is split and shown on two opposite sides of the map as if separated by 20,000 miles.

That last step required more work than is immediately apparent, however. It took Fuller two years of experimenting to find an orientation in which all twelve icosahedral vertices land in the ocean —an essential requirement if land masses are not to be ripped apart. Observe in Figure 16-1 that many of the vertices are extremely close to shore. One can imagine the frustrating task of searching for twelve water locations; moving a vertex away from land on one side of the globe would instantly result in a number of vertices bumping into land somewhere else. Contemplating the five or six angular gaps which are precariously close to land masses, one suspects that Fuller's final arrangement may be a unique solution to the problem.

The Dymaxion Map, awarded U.S. Patent 2,393,676 in 1946, is an unprecedented cartographic accomplishment, which was made possible by a straightforward application of geometry. This map is therefore another superb example of the design-science approach. Fuller considered the problem *outside* the context of traditional

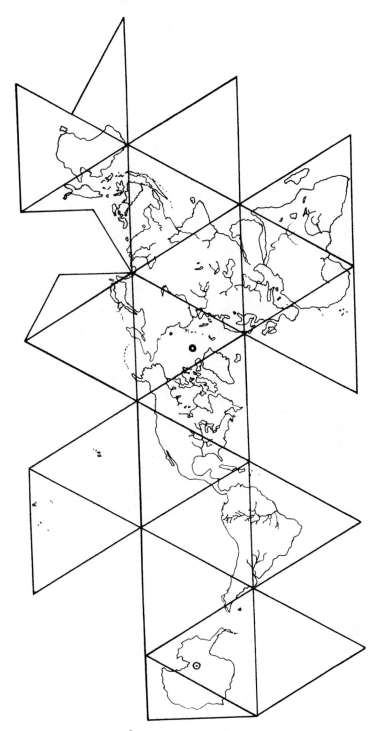

Fig. 16-1. Dymaxion Map^{T}, used with permission of the Buckminster Fuller Institute. (See Appendix C for more information about the Institute.)

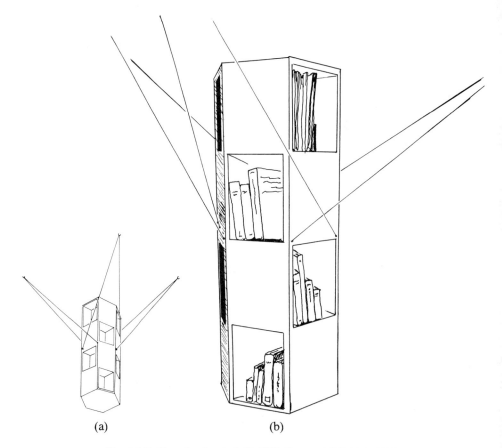

Fig. 16-2. Hanging bookshelf, U.S. Patent 4,377,114 (1983).

map-making; rather than attempting to work with and refine history's previous best solution, he started over. He sets the example of considering a design problem as a whole system.

Suspended Storage Systems

A third synergetic design application is Fuller's 1983 hanging bookshelf. Conceived as a space-saving storage device, the narrow shelving unit is suspended from the ceiling by six wires. The original design was a tall hexagonal column of wood, with compartments on all sides creating omnidirectional access, as the unit is suspended rather than leaning against the wall (Fig. 16-2). Hanging by its few cables, the column appears ready to swing back and forth at the slightest push from any direction. People would approach this structure—on display for a while in a Philadelphia bookstore—and touch it hesitantly and gently, only to be astonished by its resistance. The more daring will ultimately lean against the column with all their weight and discover that it does not move. Just as for the tensegrity sculptures, this rigidity is especially surprising because we do not expect "delicate" tension elements to be capable of the same strength as "solid" columns; however, even those who anticipate the great capability of tension are caught off guard by this display. In spite of recognition that thin cables can support a massive block of wood, the fact that the shelf does not swing even slightly out of place remains astonishing.

This invention utilizes one geometric principle: Fuller's "twelve degrees of freedom." Twelve vectors, or independent forces (six positive, six negative), must be applied to a body in space to completely restrict its mobility. In Chapter 7 we looked at the application of this inherent spatial characteristic to a bicycle wheel, and observed that a minimum of twelve spokes was required to rigidly restrain its hub. Fuller's bookshelf presents a similar design problem; why then is it anchored with only six "spokes"? The answer lies in the massiveness of this gigantic "hub"; the hexagonal column is heavy enough to pull firmly against the six wires and remove the remaining, or negative, six degrees of freedom. The cables are symmetrically arranged—as are the six spokes anchoring one half of our hypothetical bicycle hub—but in this case gravity takes care of the other six spokes, pulling in the opposite direction from each cable by taking advantage of a heavy object's considerable attraction to the earth. If the column were made of lightweight plastic, the design would not work; six degrees of freedom would still be unaccounted for.

This design represents a remarkably simple application of a synergetic principle, and produces a startling piece of furniture, the major significance of which—if saving floor space is not a consideration—may be educational. It is a profoundly reorienting experience to *feel* the precariously suspended shelf's refusal to budge.

The above examples suggest that design science can be considered a science of spatial order. As such, this study is necessarily comprehensive; space is everywhere. We also learn that invention does not spring fully formed out of principle, but rather requires a little work. First a need is ascertained, as for example for an efficient shelter system. Then, relevant principles are gathered. From the jumble of known truths, one or two might apply to a problem. The next step is to pull them in, experiment, twist them around, and not give up: seek that innovative application of an age-old principle.

More with Less

The three designs described above were chosen as straightforward examples of design science. However, Fuller's main purpose is to call our attention to an invisible design revolution already taking place, to inspire our active participation in guiding this progression in preferred ways. He points out fantastic technological advances, such as new communications satellites, each one weighting a quarter-ton and outperforming 175,000 tons of transoceanic copper cable.[3] Anyone who remembers the shaky transatlantic telephone connections of the past can appreciate the qualitative improvement as well. Similarly, ever stronger metal alloys enhance humankind's structural capability. The average person is not aware of this metallurgical revolution, says Bucky, because it cannot be seen. An invisible reality is quietly taking over, accomplishing so much more with so much less material and other resources, that the logical extension is sufficient and sustainable life support for all humanity.

To further explain this potential, Fuller discovers that he is forced to redefine "wealth." Too long associated with money and other tangible and limited resources, wealth is actually the organized capacity of society to apply its resources to take care of lives. A computer is not worth much in terms of its content of precious materials; its value is in the processing of information and knowledge. Wealth involves energy and knowledge; the former is neither created nor destroyed, and the latter is constantly increasing. Therefore, humanity's true wealth is constantly increasing and has no inherent upper limit. There is a fixed amount of gold in the world, but the currency of our emerging era is knowledge and its creative application.

Finally, Fuller points out that Thomas Malthus could not have forseen this technological revolution. His epochal conclusion in 1805 that population increases geometrically while its resources—ability to feed, clothe, and house itself—expand arithmetically at best, is now obsolete. Malthus did not anticipate the phenomenon of more effective performance using less resources. His declaration predates refrigeration, let alone the information and communications revolution, emphasizes Bucky, and yet humanity's social and economic institutions are still based on the assumption of fundamental scarcity. Malthusian thinking has controlled human affairs for so long that we have mistaken it for absolute truth. The only barrier to a successfully sustainable planet is ignorance, Bucky declares. Fundamental scarcity is a remnant of the dark ages.

The essential message of Fuller's design science is that human beings have access to the design laws of Universe, and a responsibility to use the extraordinary phenomenon of mind to discover and apply such principles. Our function is problem-solving. Synergetics, the discipline behind Fuller's more-with-less philosophy, above all encourages us to experiment. This material is superbly suited to nurture and enhance creativity, demanding both numerical rigor and intuitive leaps. The systematic study of spatial complexities is still young, and its significance and utility as yet undeveloped. The future is wide open, but we must probe and step beyond our fragile equilibrium, if Fuller's vision is to be tested.

A design science revolution is imperative.

Appendix A: Trigonometric Calculations

Finding Interior Angles of Polygons

Pentagon

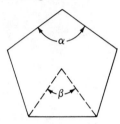

$$360° \div 5 = \beta$$
$$72° = \beta$$
$$\beta + 2\left(\tfrac{1}{2}\alpha\right) = 180°$$
$$\alpha = 180° - \beta$$
$$\alpha = 108°$$

Hexagon

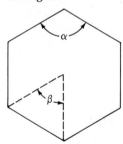

$$360° \div 6 = \beta = 60°$$
$$\beta + \alpha = 180°$$
$$\alpha = 120°$$

Heptagon

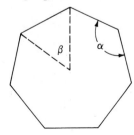

$$360° \div 7 = \beta = 51.43°$$
$$\alpha = 180° - 51.43°$$
$$\alpha = 128.57°$$

Chord Factors

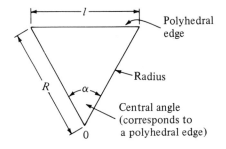

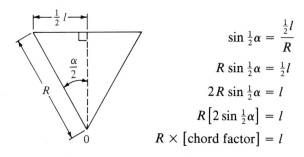

$$\sin \tfrac{1}{2}\alpha = \frac{\tfrac{1}{2}l}{R}$$

$$R \sin \tfrac{1}{2}\alpha = \tfrac{1}{2}l$$

$$2R \sin \tfrac{1}{2}\alpha = l$$

$$R\left[2 \sin \tfrac{1}{2}\alpha\right] = l$$

$$R \times [\text{chord factor}] = l$$

For every central angle α, there is a numerical value called a "chord factor," generated by the equation $[2 \sin \tfrac{1}{2}\alpha]$, which can be multiplied by the desired radius of a polyhedral system to calculate the exact length of the chord (edge) subtended by α.

Appendix B: Volume Calculations for Three Prime Structural Systems

Tetrahedron

Using traditional formula:

$$V = \tfrac{1}{3}A_b H,$$

where A_b = area of base and H = height.

(1) Area of base:

$$A_b = \frac{Bh}{2} = \frac{1 \times \sin 60^\circ}{2}$$

$$= \frac{1(0.8660)}{2}$$

$$= 0.4330.$$

(2) Height of tetrahedron:

$$H^2 + (0.5774)^2 = 1^2,$$

$$H = 0.8165.$$

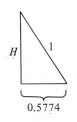

(3) Volume:

$$V = \frac{A_b \times H}{3} = 0.11785.$$

Octahedron

Two square-based pyramids:

$$\left[\frac{A_b H}{3}\right] \times 2 = V.$$

(1) Area of base:

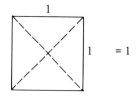   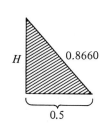

(2) Height of pyramid:

$$H^2 + (0.5)^2 = (0.8660)^2,$$
$$H = 0.7071.$$

(3) Volume of half octahedron:

$$\frac{A_b \times H}{3} = \frac{(0.7071)(1)}{3} = 0.2357.$$

(4) Volume of octahedron:

$$2 \times 0.2357 = 0.47140.$$

Icosahedron

Twenty pyramids with equilateral bases and side-edge length equal to the icosahedron radius.

(1) Pentagonal cross-section:

$$x = 2[\sin 54° \times 1]$$
$$= 1.61803$$

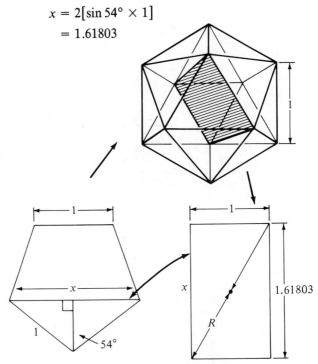

(2) Radius R:

$$(2R)^2 = (1.61803)^2 + (1)^2,$$
$$R = 0.95106.$$

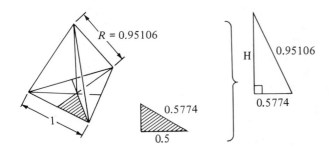

(3) Pyramid height H:

$$(0.95106)^2 = (.5774)^2 + H^2,$$
$$H = 0.7557.$$

(4) Area of pyramid base is equal to area of base of regular tetrahedron:

$$A_b = \frac{1(0.866)}{2} = 0.4330.$$

(5) Pyramid volume:

$$\tfrac{1}{3}(A_b H) = 0.1091.$$

(6) Volume of icosahedron:

$$20 \times (0.1091) = 2.1817.$$

Appendix C: Sources of Additional Information

All books, unless noted by an asterisk, are available from the Buckminster Fuller Institute, 1743 S. La Cienega Boulevard, Los Angeles, CA 90035. A number of versions of the Dymaxion Map can also be obtained; write the Institute for a complete list.

General

Some of the other books by R. Buckminster Fuller (in addition to those listed in the bibliography) recommended for further reading:

Operating Manual for Spaceship Earth (Carbondale, Ill.: Southern Illinois University Press, 1969).
No More Secondhand God (Carbondale, Ill.: Southern Illinois University Press, 1969). Poems and essays.

Biographical Material

Robert Snyder, *Buckminster Fuller: An Autographical Monologue Scenario*, (New York: St. Martin's Press, 1980). A great introduction, written by Fuller's son-in-law using Fuller's own words, with hundreds of photographs from the Fuller archives.
E. J. Applewhite, *Cosmic Fishing* (New York: Macmillan, 1977). Especially good for its description of the collaboration between Bucky and his long-time friend E. J. Applewhite in the process of writing *Synergetics*.
Hugh Kenner, *BUCKY: A Guided Tour of Buckminster Fuller* (New York: William Morrow, 1973).

Geodesic Mathematics

Hugh Kenner, *Geodesic Math: And How to Use It* (Berkeley, University of California Press, 1976).

World Game

"50 Years of Design Science Revolution and the World Game," historical
   documentation (articles, clippings) with commentary by Fuller, 1969.
   Available only through the Buckminster Fuller Institute.
Medard Gabel, *Earth, Energy and Everyone* (Garden City, N.Y.:
   Doubleday, 1975).
Medard Gabel, *HO-ping: Food For Everyone* (Garden City, N.Y.:
   Doubleday, 1976).

Write for newsletter, information about seminars and workshops, and list of
available material to The World Game, University City Science Center,
3508 Market Street #214, Philadelphia, PA 19104.

Appendix D: Special Properties of the Tetrahedron

(1) Minimum system: the tetrahedron is the first case of insideness and outsideness.

(2) The regular tetrahedron fits inside the cube, with its edges providing the diagonals across the cube's six faces, and thereby supplying the six supporting struts needed to stabilize the otherwise unstable cube. Furthermore, two intersecting regular tetrahedra outline all eight vertices of the cube.

(3) The tetrahedron is unique in being its own dual.

(4) The six edges of the regular tetrahedron are parallel to the six intersecting vectors that define the vector equilibrium.

(5) Similarly, the four faces of the regular tetrahedron are the same four planes of symmetry inherent in the vector equilibrium and in cubic closepacking of spheres. The tetrahedron is thus at the root of an omnisymmetrical space-filling vector matrix, or isotropic vector matrix.

(6) When the volume of a tetrahedron is specified as one unit, other ordered polyhedra are found to have precise whole-number volume ratios, as opposed to the cumbersome and often irrational quantities generated by employing the cube as the unit of volume. Furthermore, the tetrahedron has the most surface area per unit of volume.

(7) Of all polyhedra, the tetrahedron has the greatest resistance to an applied load. It is the only system that cannot "dimple"; reacting to an external force, a tetrahedron must either remain unchanged or turn completely "inside out."

(8) The surface angles of the tetrahedron add up to 720 degrees, which is the "angular takeout" inherent in all closed systems.

(9) The tetrahedron is the starting point, or "whole system," in Fuller's "Cosmic Hierarchy," and as such contains the axes of symmetry that characterize all the polyhedra of the isotropic vector matrix, or face-centered cubic symmetry in crystallography.

(10) Packing spheres together requires a minimum of four balls, to produce a stable arrangment, automatically forming a regular tetrahedron. The centers of the four spheres define the tetrahedral vertices. In Fuller's words, "four balls lock."

(11) It has been demonstrated that many unstable polyhedra can be folded into tetrahedra, as in the jitterbug transformation.

(12) Fuller refers to the six edges of a tetrahedron as one "quantum" of structure, because the number of edges in regular, semiregular, and high-frequency geodesic polyhedra is always a multiple of six.

Appendix E: Glossary

Includes specific terminology coined and/or used in an unusual way by Buckminster Fuller (in italics) as well as terms from conventional mathematics and geometry that may be unfamiliar.

A-module: Asymmetrical tetrahedron, which encompasses one twenty-fourth of the regular tetrahedron.

acute angle: Angle less than 90 degrees.

angle: Formed by two lines (or vectors) diverging from a common crossing. An angle is necessarily independent of size.

angular topology: Fuller's term, intended to describe a principal function of synergetics, that is, description of all structure and pattern by variation of only two variables: angle and frequency. (See "frequency.")

arc: A segment of a curve.

B-module: Asymmetrical tetrahedron, which is the result of subtracting an A module from one forty-eighth of the regular octahedron.

complementarity, or *inherent complementarity of Universe:* Necessary coexistence, inseparable pairs. Fuller emphasizes that Universe consists of complementary teams, such as concave–convex, tension–compression, positive–negative, and male–female, and concludes that "unity is inherently plural."

concave: Curved toward the observer, such as the surface of a sphere, or other enclosure, as seen from the inside.

convex: Curved away from the observer, such as the surface of a sphere as seen from the outside.

coupler: Semisymmetrical octahedron, which consists of sixteen "A-Modules" and eight "B-Modules," or eight "Mites."

degrees of freedom: This term is used by Fuller to mean the number of independent forces necessary to completely restrain a body in space, and by Loeb in reference to the overall stability of systems.

design: Deliberate ordering of components.

design science: See Chapter 16.

dimension: See Chapter 6. Used by Fuller to include spatial extent, orders of complexity, and distinct facets of symmetry.

dimpling: Yielding inwardly to produce local indent in structural system, caving-in.

Dymaxion: Fuller's trademark word, created by Marshall Fields Department Store in the late 1920s, for use in promotion of exhibit featuring Fuller's revolutionary house design. Also used by Fuller to mean "doing more with less."

edge: Geometry term: connection between points (or vertices) and boundary between two faces, line. Fuller prefers to substitute "vector" or else refer to the specific material of construction, such as toothpick or straw.

energy event: Fuller's substitute for geometric term "point." Also descriptive term for natural phenomena, discrete constituents of Universe. Replaces "obsolete" vocabulary such as "solid," "point," "thing," etc.

ephemeralization: Doing more with less, via design science and technological invention.

face: Geometric term: polyhedral window, polygon, area.

finite accounting system: Describes concept of physical reality consisting, on some level, of discrete indivisible particles, as opposed to continuous surfaces or masses. See Chapter 2.

frequency: Used by Fuller to specify length and size in general. The intention is to employ a more precisely descriptive term for both geometric systems and events in nature, than provided by specific units of measurement. Fuller points out that frequency never relates to the quantity "one," for it necessarily involves a plurality of experiences.

high-frequency energy event: Describes most tangible structures, which might be popularly thought of as "solids." A good example to illustrate the concept of "high-frequency energy event" through a visible image is found in the white foam of breaking ocean waves. Upon closer inspection, the apparent continuum of whiteness is a result of an enormous number of tiny clear bubbles, which appear continuous because of their close proximity. This punctuated consistency is analogous to all matter, although most examples are not visible to the naked eye.

frequency of modular subdivision: Number of subdivisions per edge, in a polyhedron, or number of discrete subdivisions per module in any system.

generalized principle: Rule that holds true without exception. Eternal law of nature. See Chapter 1.

great circle: A circle on the surface of a sphere, which lies in a plane intersecting the center of that sphere. Equatorial ring.

great-circle arc: Segment of great circle.

Greater Intellectual Integrity of Universe: Sum total of all "generalized principles," complex "unknowable" totality of Universe. Divine intellect. Perfection of eternally regenerative pattern integrities of Universe.

in, out: Seen by Fuller as appropriate replacements for up, down, to describe the directions toward the earth's center and away from the earth's surface, respectively.

inherent complementarity of Universe: See "complementarity."

interior angle of polygon: Angle between two edges measured through the inside.

interprecessing: See *precession.*

intertransformability: Phenomenon of significant relationships between systems, allowing transformations from one to another. Applies to both polyhedra and natural structures. Examples include shared symmetry among polyhedra and common constituents arranged differently to produce different substances.

isotropic vector matrix, IVM: Space-filling array of unit-length vectors, in which all vectors are identically situated. The "omnisymmetrical" matrix consists of an indefinite expanse of alternating tetrahedra and octahedra, with 60-degree angles between adjacent vectors. This conceptual IVM framework can be actualized with building materials to create an Octet Truss, which is an extremely efficient and lightweight architectural space frame.

jitterbug transformation: Transformation of a cuboctahedron model, in which flexible joints allow the unstable polyhedron to contract in a radially symmetrical manner, and thereby take on the shape of various other polyhedral systems.

Mite: Two mirror-image "A-modules" and one "B-module" combined to create an asymmetrical tetrahedron, in which three right angles surround one vertex. Trirectangular tetrahedron.

nature's coordinate system: The mathematically expressible system that governs the coordination of both physical and metaphysical phenomena. Set of generalizations about the way systems are structured and able to cohere over time. Interplay of the principles describing spatial complexity with the requirements of minimum energy in the organization of natural structures.

nest: Valley, or local indentation, between adjacent closepacked spheres.

net: Planar array of adjacent polygons which can be folded along shared edges and closed together to create a specific polyhedron.

obtuse angle: Angle greater than 90 degrees.

omni-accommodative: Able to accommodate all spatial directions, or model all transformations; omnidirectional.

omni-interaccommodative: Describes cooperative relationship between noncontradictory principles or evidence, which are thus more significant considered together than separately.

omnisymmetrical: Symmetry in all spatial directions.

pattern integrity: Reliable or consistent arrangement of "energy events" (or constituent parts) in dynamically regenerative system. A pattern with structural integrity, that is, a pattern that coheres for some period of time.

precession: Two or more systems in motion with respect to each other involving 90-degree turn. In addition to its meaning in physics—describing a complex motion of a rotating body in response to an applied torque—Fuller employs this word (as well as his own "interprecessing") to refer to two geometrical systems which, oriented perpendicularly to each other, reveal a new system or geometric relationship.

quanta: Used by Fuller to mean indivisible discrete units, limit-case particles, isolated energy events. See *finite accounting system.*

regular polyhedron: Polyhedron composed exclusively of one type of polygonal face meeting at identical vertices.

right isosceles triangle: Triangle which includes two equal edges (or arc lengths) with a right-angle between them. Planar version has angles of 45°–45°–90°; one particular spherical version has angles of 60°–60°–90°.

similar: Geometrical term: having the same shape, but not necessarily the same size.

size: Dimension, extent. Relates to actual constructs, or "special-case" systems.

Spaceship Earth: Coined by Fuller to convey a sense of a finite, whole system planet, in which the lives of all human beings (or passengers) are interrelated. The idea is to encourage thinking of Earth as a single system with a common interest in successful survival.

special-case: Relates to specific example rather than generalized system. Used especially to refer to specific manifestation of generalized concept.

spheric: Rhombic dodecahedron. Fuller's term, derived from the fact that the rhombic dodecahedron is the domain of each sphere in closest packing.

spherical triangle: A curved area bounded by three connected great-circle arcs. The result of interconnecting three points on the surface of a spherical system.

structural system: Triangulated system.

supplementary angles: Angles that add up to 180 degrees.

synergy: Behavior of whole systems not predictable from the behavior of separate parts.

system: Four or more interrelated "events."

tune-in-ability: Possibility of isolating or focusing on specific phenomena as independent systems despite interrelatedness of such phenomena to many other systems. (See Chapter 3.) Boundaries can be constantly redefined, according to the particular goal of an investigation. Also refers to limits of resolution in defining systems.

Universe: Paraphrasing Fuller, Universe is the aggregate of all experience. (See Chapter 1.) The role of the observers, or humanity, is an essential component of the definition, for awareness is a prerequisite to defining and understanding. Experience consists of dynamic, regenerative patterns of energy—perhaps "omnidynamic" would be an appropriate term.

-valent: Number of connections or elements, such as number of edges meeting at a vertex, or number of sides of a polygonal face; e.g., "four-valent" vertex of an octahedron, and "three-valent" face (triangle) of the same, or three-valent vertex (and face) of a tetrahedron.

valving: Deliberate channeling of energy and resources in preferred ways.

vector: Represents energy event, consists of magnitude and direction, represented on paper by an arrow with specific length (or frequency) and angular orientation, and used by Fuller instead of "edges" in polyhedra. Vectors also represent *relationships* between energy events.

vector equilibrium: Fuller's term for the cuboctahedron, in recognition of its equal radial and edge lengths.

vertex: Geometrical term: crossing, convergence of lines or edges, joint, point, polyhedral corner.

wealth: Organized capacity of society to apply generalized principles toward present and future life support.

whole number: In mathematics, positive integer, without fractional part.

Bibliography

E. J. Applewhite, *Cosmic Fishing* (New York: Macmillan, 1977).

Buckminster Fuller, *Synergetics: The Geometry of Thinking* (New York: Macmillan, 1975).

Buckminster Fuller, *Synergetics 2: Further Explorations in The Geometry of Thinking* (New York: Macmillan, 1979).

Buckminster Fuller, *Utopia or Oblivion* (London: Allan Lane, The Penguin Press, 1979).

Buckminster Fuller, *Critical Path* (New York: St. Martin's Press, 1981).

Buckminster Fuller, *Inventions* (New York: St. Martin's Press, 1983).

Buckminster Fuller, *Intuition* (Garden City, N.Y.: Anchor Press/Doubleday, 1973).

Buckminster Fuller and Robert Marks, *The Dymaxion World of Buckminster Fuller* (Garden City, N.Y.: Anchor Press/Doubleday, 1973).

Hugh Kenner, *Bucky: A Guided Tour of Buckminster Fuller* (New York: William Morrow, 1973).

Arthur L. Loeb, *Space Structures: Their Harmony and Counterpoint* (Reading, Mass.: Addison-Wesley, Advanced Book Program, 1976).

Peter Pearce, *Structure in Nature Is a Strategy for Design* (Cambridge, Mass.: MIT Press, 1978).

Notes

Chapter 1

[1]Buckminster Fuller, *Synergetics: The Geometry of Thinking* (New York: Macmillan Publishing, 1976), p. 30, section 216.03: "Comprehension of conceptual mathematics and the *return to modelability* are among the most critical factors governing humanity's epochal transition from bumblebee-like self's honey-seeking preoccupation into the realistic prospect of a spontaneously coordinate planetary society." (My italics—A.E.) As stated in the introductory "Note to Readers," quotations from *Synergetics* will hereafter simply be followed by a numerical section reference in parenthesis. Quotation from *Synergetics 2: Further Explorations in the Geometry of Thinking* will be followed by the section number and the letter "b."

[2]These statements were included in many of Fuller's lectures. I am quoting from memory and from personal notes made over the years. Hereafter, assume this to be the source if not otherwise referenced.

[3]See later chapters, especially Chapter 3 for explanation of alloys, and Chapter 16 for discussion of Malthus.

[4]See "Spaceship Earth" in the Glossary (Appendix E).

[5]Sir Arthur Eddington, (1882–1944), English astronomer and physicist.

[6]Quoted from epic videotaped lecture session, recorded in January 1975, in Philadelphia, over a ten-day period. The total length of these lectures, which are part of Fuller's archives, now located in the Buckminster Fuller Institute in Los Angeles, is 43 hours. (Hereafter referred to as "43-hour videotape.")

[7]See e.g. *Powers of Ten*, by Philip and Phyllis Morrison and The Office of Charles and Ray Eames (New York: Freeman, Scientific American Books, 1982).

[8]See Appendix C for biographical references.

[9]Arthur L. Loeb, *Space Structures* (Reading, MA: Addison-Wesley, Advanced Book Program, 1976), p. xvii.

[10]$F = GMm/r^2$: The attractive force due to gravity between two objects is equal to the "gravitational constant" (G) times the product of their masses and divided by the separation-distance raised to the second power.

Chapter 2

[1]See Glossary, "whole number."

[2] See Appendix C.

[3] It must be noted that tetrahedra cannot fit together face to face to form the larger tetrahedra shown, but rather must alternate with octahedra, as will be explored at length in later chapters (8, 9, 10, 12, 13). The values given for the volume of each tetrahedron are based on using a unit-length tetrahedron as one unit of volume, in exactly the same manner that a unit-length cube is conventionally employed. Despite the tetrahedron's inability to fill space, the relative volumes of tetrahedra of increasing size are identical to those exhibited by cubes of increasing size.

Chapter 3

[1] See Loeb's preface to Fuller's *Synergetics*: "In rejecting the predigested, Buckminster Fuller has had to discover the world by himself" (p. xv).

[2] Buckminster Fuller, *Intuition* (Garden City, N.Y.: Anchor Press/ Doubleday, 1973), p. 39.

[3] See Glossary, "spherical triangle," "concave," "convex."

[4] 43-hour videotape.

[5] Quoted several times by Fuller, in 43-hour videotape.

[6] Hugh Kenner, *Bucky: A Guided Tour of Buckminster Fuller* (New York: William Morrow, 1973), p. 113. See Glossary (Appendix E) for definition of "Dymaxion."

Chapter 4

[1] "Contribution to Synergetics" is a 52-page supplement by Arthur Loeb included in Fuller's *Synergetics*, pp. 821–876.

[2] See Glossary, "-valent".

[3] Arthur Loeb, *Space Structures*, Chapter 6.

[4] Loeb, p. 11.

[5] Chapter 5 explains why "structural system" implies triangulated system.

[6] Loeb, p. 40.

[7] Loeb, p. 63.

[8] See Glossary, "acute" and "obtuse."

Chapter 5

[1] The best renditions of the precession sequence are found in videotaped lectures, because Fuller's gestures are as important as his words; the 43-hour videotape contains an especially good version. Original written document was published in *Fortune*, May, 1940. Fuller wrote the two-page piece in response to a request (or challenge!) from the Sperry Gyroscope company.

[2]See Glossary, "Greater Intellectual Integrity of Universe."

[3]Loeb, pp. 29–30. The derivation of the equation $3V - E = 6$, which is thought to have originated with Maxwell, can be derived from degree-of-freedom analysis of molecular spectra, and was applied by Loeb to polyhedral systems.

Chapter 6

[1]See Glossary, "special case."

[2]43-hour videotape.

[3]Kenner, pp. 129–131.

[4]Appendix A, "chord factors."

Chapter 7

[1]William Morris, Editor, *The American Heritage Dictionary* (New York: American Heritage Publishing Co., 1970), p. 442.

[2]43-hour videotape.

[3]It must be noted however that if the cuboctahedron is sliced in half and one "hemisphere" is rotated 60 degrees with respect to the other, the resulting "twist cuboctahedron" (Fig. 7-6b) maintains the radial–circumferential equivalence. With its asymmetrical arrangement of faces, however, this shape is not similarly suited to model equilibrium. The desired balance of vectors is therefore achieved through the straight cuboctahedron (Fig. 7-6a).

[4]Amy Edmondson, "The Minimal Tensegrity Wheel," Thesis for completion of B.A. degree requirements, Harvard University, 1980. Refer to VES Teaching Collection, Carpenter Center, Harvard University.

[5]Fuller's use of the term "degrees of freedom" must be distinguished from the conventional treatment of the subject which specifies that a rigid body has six degrees of freedom (three translational and three rotational) which have two directions each, thus requiring twelve unidirectional constraints.

Chapter 8

[1]N. J. A. Sloane, "The Packing of Spheres," *Scientific American* (January 1984), p. 116.

[2]Cubic packing thus corresponds to the VE (cuboctahedron), while hexagonal packing corresponds to the "twist" VE. See Footnote 3 in Chapter 7 for the difference.

[3]I. Rayment, T. S. Baker, D. L. D. Caspar, and W. T. Murakami, "Polyoma Virus Structure at 22.5 Å Resolution," *Nature*, 295 (14 January 1982), p. 110–115.

[4]Donald Caspar and Aaron Klug, "Physical Principles in the Construction of Regular Viruses," *Cold Spring Harbor Symposia on Quantitative Biology*, XXVII (1963), pp. 1–3.

Chapter 9

[1]See Glossary, "supplementary angles."

[2]A. L. Loeb, "Contribution to Synergetics," in *Synergetics: The Geometry of Thinking* (New York: Macmillan, 1976), pp. 860–875; "A Systematic Survey of Cubic Crystal Structure," *J. Solid State Chemistry*, 1 (1970), pp. 237–267.

[3]43-hour videotape.

[4]Refer to Chapter 5 and Appendix D.

Chapter 10

[1]A. L. Loeb, "Contribution to Synergetics," pp. 821–876. Volume ratios are derived by comparing geometrically similar polyhedra so that "shape constants" cancel out of the formulae. Credit for this approach belongs to Loeb and Pearsall, and it must be noted that this section (H) of Contribution to Synergetics is adapted from an article by Loeb in *The Mathematics Teacher*.

[2]Loeb, p. 836.

[3]Loeb, p. 832.

Chapter 11

[1]Loeb, p. 829.

[2]A. L. Loeb, "Addendum to Contribution to Synergetics," published in *Synergetics 2: Further Explorations in the Geometry of Thinking*, (New York: Macmillan Publishing, 1979), pp. 473–476.

[3]Dennis Dreher of Bethel, Maine designed an omnidirectional hinging joint, which allows the necessary twisting of adjacent VE triangles in the jitterbug. This joint can be seen in Photograph 11-1.

Chapter 12

[1]"Bohr, Niels" *Encyclopaedia Britannica*, 1971.

[2]Isaac Asimov, *Asimov's Guide to Science* (New York: Basic Books, 1972), pp. 334–5.

[3]Loeb, "Contribution to Synergetics," pp. 836–847.

[4]Loeb, "Coda," *Space Structures*, pp. 147–162.

Chapter 13

[1] Discovered independently by Loeb and called "moduledra."

[2] Loeb, "Contribution to Synergetics" pp. 847–855;

[3] *Space Structures*, "Coda," pp. 147–162.

Chapter 14

[1] See Appendix E, "right isosceles triangles."

[2] Fuller, *Synergetics*, Figure 455.20, p. 178.

[3] Fuller, *Synergetics*, Figure 458.12, p. 189.

[4] See Appendix E and Chapter 15.

Chapter 15

[1] Buckminster Fuller, *Critical Path* (New York: St. Martins Press, 1981), p. 13.

[2] This fact was demonstrated in the "tensegrity" bicycle wheel experiment mentioned in Chapter 7. In tests to determine the minimum number of tension (Dacron string) "spokes" required to stabilize its hub, the wheel's eventual structural failure originated with buckling of its rim. Long arc spans (which were a consequence of the low number of radial spokes) were too thin to withstand the compression force created by loading the hub and transmitted to the rim through the tension spokes. The usual arrangement, which consists of a fairly large number of spokes (36 or more), therefore turns out to be advantageous; despite the fact that there are many more spokes than necessary to restrain the hub, this large number does serve to subdivide the otherwise vulnerable arc segments of the compression-element rim. See also Edmondson, "The Minimal Tensegrity Wheel."

[3] Recalling the principle of angular topology in Chapter 6, we also know that the "angular takeout" is 720 degrees. That is, if we subtract the sum of the surface angles at each vertex of a convex polyhedron from 360 degrees, the sum of all these differences will be exactly 720 degrees.

[4] Ernst Haeckel, *Art Forms in Nature* (New York: Dover, 1974), plate 1 (Radiolaria) and plate 5 (calcareous sponges).

[5] See Chapter 8, notes 3 and 4.

[6] Note that the engineering term "pure axial force" is thus a convenient simplification, which is effective in terms of structural analysis, rather than an accurate scientific description. A strut which is to carry either axial compression or tension can be significantly lighter than one which is subject to bending or torque.

[7] Buckminster Fuller, "Tensegrity," *Creative Science and Technology*, IV, No. 3 (January-February, 1981), p. 11.

[8]*Synergetics*, p. 354.

[9]See note 2 in Chapter 1.

[10]This very general summary is based on Dr. Donald Ingber's paper, "Cells as Tensegrity Structures: Architectural Regulation of Histodifferentiation by Physical Forces Transduced over Basement Membrane," published as a chapter in *Gene Expression During Normal and Malignant Differentiation*, (L. C. Anderson, C. G. Gahmberg, and P. Ekblom, eds.; Orlando, Fla.: Academic Press, 1985), pp. 13–22.

[11]Taken from a letter from Ingber to Buckminster Fuller, April 5, 1983.

Chapter 16

[1]This approach characterizes Fuller's "World Game" studies. See Appendix C for sources of information about this research.

[2]Buckminster Fuller and Shoji Sadao, Cartographers; Copyright R. Buckminster Fuller, 1954. "Dymaxion Map" is a trademark of the Buckminster Fuller Institute.

[3]Fuller, *Critical Path*, p. xxiii.

Index

(**Bold print** indicates page number which includes illustration of entry.)

DISCARD